Instructions for Borrowing a Planet:

Rediscovering Wonder, Humility, and Our Place in the Living World

Mark Donnelly, PhD.

RPSS - ROCK PAPER SAFETY SCISSORS PUBLISHING

drmaddog@hotmail.com

Instructions for Borrowing A Planet

Perfect Bound ISBN: 978-1-956688-85-6

Printed in the United States of America

10 9 8 7 6 5 4 3 2 1

RPSS Publishing - Buffalo, New York

Dedication:

To the generations we will never meet.

May we borrow this remarkable planet with gratitude, wonder, and humility, and return it just a little kinder, a little healthier, and a little more beautiful than the way we found it.

Table of Contents

BORROWER'S AGREEMENT

Instructions for Borrowing a Planet

Borrower: ______________________________

Date Received:

The day you were born.

Return Date:

Unknown.

Item Borrowed:

One (1) living planet, including but not limited to:

- Oceans
- Mountains
- Rivers
- Forests
- Prairies
- Deserts
- Clouds
- Rain
- Sunrises
- Sunsets
- Approximately eight million known species, plus a great many we haven't met yet
- One astonishing atmosphere
- An unreasonable number of stars visible on clear nights
- Unlimited opportunities to be surprised

Terms and Conditions

By accepting temporary residence on this remarkable world, you agree to the following:

1. Remember that you are borrowing, not owning.

The trees were here before you.

The rivers learned their paths without your advice.

The mountains have been practicing patience for millions of years.

You are joining an ancient neighborhood.
Please introduce yourself kindly.

2. Look closely.

Notice birds that sing before your alarm clock.

Watch ants carrying burdens that seem mathematically impossible.

Learn the names of at least a few trees, flowers, and clouds.

Curiosity is considered proper maintenance.

3. Leave more than footprints.

Plant something.

Teach someone.

Restore a trail.

Pick up litter that isn't yours.

Offer encouragement as generously as shade.

Small acts accumulate into large legacies.

4. Take only what you need.

The Earth has always been generous.

Try not to confuse generosity with permission to waste.

5. Share generously.

Knowledge.

Wonder.

Seeds.

Stories.

Picnic blankets.

Homemade pie.

The last one isn't required, but it has an excellent track record.

6. Speak well of your neighbors.

This includes the human ones.

It also includes bees, mushrooms, chickadees, moss, earthworms, old oak trees, and that slightly opinionated squirrel who seems convinced your bird feeder belongs to him.

You don't have to understand every neighbor.

It helps if you learn a few names.

7. Practice gratitude often.

Clean water deserves thanks.

So does fresh air.

So does fertile soil.

So does the stranger who planted a tree whose shade you enjoy decades later.

Most of life's greatest gifts arrive already prepared by someone–or something–you will never meet.

8. Expect occasional inconvenience.

There will be mosquitoes.

Mud.

Wind.

Unexpected rain.

Leaves to rake.

Birds that express surprisingly strong opinions before sunrise.

These are features, not defects.

9. Resist becoming too busy to notice miracles.

Fireflies still appear.

The moon still rises.

Seeds still split open in the dark.

Forests quietly invent spring every year.

Don't miss them while answering emails. The emails can wait.

Spring has a tighter schedule.

10. Return the planet in better condition than you found it whenever possible.

You may not solve every environmental problem. Few people do.

But every kindness leaves evidence.

Every tree planted.

Every stream restored.

Every child taught to wonder.

Every creature protected.

Every thoughtful choice becomes part of the inheritance left for those who arrive after you.

Fine Print

This agreement is renewed every morning.

Failure is expected. Perfection is not.

Wonder is strongly encouraged.

Curiosity is renewable.

Humility improves with practice.

Laughter is compatible with wisdom.

Hope requires participation.

Love grows wherever attention is consistently applied.

A Final Reminder

You will not be remembered because you owned a small corner of the Earth.

You will be remembered because, for a little while, the Earth was healthier, kinder, more beautiful, or more hopeful because you passed through it.

Thank you for borrowing responsibly.

Please enjoy your visit.

Introduction

We Were Never the Owners

Somewhere along the way, we began behaving as though the Earth belonged to us.

It's an understandable misunderstanding.

The mailbox has our name on it.

The mortgage says we own the house.

The deed says we own the land.

The tax bill certainly agrees.

But the oak tree in the backyard has never seen the paperwork.

Neither have the robins.

Or the earthworms.

Or the fox that occasionally wanders through just before dawn, looking mildly disappointed in the neighborhood's rabbit population.

Nature has always ignored our filing system.

That, I have come to believe, is one of its more charming qualities.

The older I get, the more suspicious I become of the word ownership.

I own a favorite coffee mug.

Until gravity reminds me that I merely borrowed it for a while.

I owned a full head of hair.

Apparently, that was a temporary arrangement.

I once owned knees that could climb mountains without filing a formal complaint. Those, too, have entered a different phase of their careers.

Perhaps life has been trying to teach us something all along.

Maybe ownership is simply borrowing with unreasonable confidence.

The same may be true of this remarkable blue planet.

We arrive with empty hands.

We leave with empty hands.

Between those two moments, we spend a few remarkable decades walking through forests that were growing long before we were born, drinking water that has circled the globe since before human history, breathing oxygen generously supplied by organisms that never send invoices.

Everything essential has already been quietly working without us.

That realization is not discouraging.

It is wonderfully freeing.

Because if we are borrowers instead of owners, our responsibilities become beautifully simple.

Borrow things carefully.

Use them gratefully.

Return them in better condition whenever possible.

It sounds like advice from a good neighbor because, in many ways, it is.

For most of my life, I thought environmentalism was primarily about solving enormous problems.

Climate.

Pollution.

Deforestation.

Species loss.

Those are, without question, enormous problems.

They deserve serious science, thoughtful policy, engineering, economics, and determined people willing to devote their lives to difficult work.

This book is not a replacement for any of that.

It is an invitation to begin somewhere earlier.

Much earlier.

Before policy comes perception.

Before stewardship comes affection.

Before affection comes attention.

And attention begins with something so ordinary that we almost overlook it.

Curiosity.

Children understand this instinctively.

Give a five-year-old a stick and twenty minutes in the woods, and you've created an afternoon.

Give the same child a puddle, and you've accidentally created a research laboratory.

Every rock deserves lifting.

Every bug deserves inspection.

Every feather raises another question.

Adults, on the other hand, have become remarkably efficient at walking past miracles.

We hurry beneath trees that manufacture breathable air without applause.

We pass birds whose migration routes stretch thousands of miles, scarcely looking up from the glowing rectangles in our hands.

We mow over dandelions without ever noticing that they are among the first spring meals for hungry pollinators.

We complain about rainy weekends while forgetting that every loaf of bread, every apple, every cup of coffee, every forest, every wildflower, every lake, and every living creature depends upon water falling from the sky with astonishing reliability.

We have become surrounded by wonders that have suffered the unfortunate fate of becoming familiar.

Familiarity is one of the greatest camouflage artists in the world.

It hides miracles in plain sight.

That is one reason I admire scientists.

The best scientists never really outgrow childhood.

They simply acquire better notebooks.

They continue asking questions long after the rest of us have become comfortable with assumptions.

Why do trees share nutrients?

How do mushrooms connect entire forests?

Why do birds sing before sunrise?

How do tiny insects pollinate crops that feed billions of people?

How does a teaspoon of healthy soil contain more living organisms than many cities contain people?

The answers are almost always stranger–and more beautiful–than the questions.

Over the past century, remarkable observers have helped us see the living world with fresh eyes.

A forester taught us to think of land as a community rather than a commodity.

A marine biologist revealed that chemicals meant to solve one problem could quietly unravel entire ecosystems.

Ecologists uncovered astonishing partnerships among plants, fungi, insects, birds, and mammals.

Botanists reminded us that gratitude can be a way of knowing.

Naturalists showed us entire universes hidden within a single square meter of forest floor.

Conservationists demonstrated that restoring a habitat sometimes begins with something as humble as planting the right flower in the right place.

Their discoveries differ in detail, but together they tell one magnificent story.

Everything is connected.

Not metaphorically.

Literally.

The oxygen you inhale this morning may have passed through a rainforest, a prairie grass, or a microscopic plankton drifting in the ocean.

The rain that falls on your roof may once have floated above Antarctica, crossed the Pacific, wandered over mountain ranges, and finally arrived at your garden carrying the memory of half a world.

The apple in your kitchen owes its existence to sunlight, fungi, bacteria, pollinating insects, healthy soil, clean water, patient growers, and seasons that somehow continue arriving on schedule despite humanity's frequent attempts to complicate matters.

Nothing truly lives alone.

Neither do we.

The older I become, the less interested I am in winning arguments about nature.

I would rather help someone fall in love with a chickadee.

Because people protect what they love.

They love what they know.

They know what they have taken the time to notice.

That simple progression may be one of the most important truths in this book.

Notice.

Appreciate.

Protect.

It sounds almost embarrassingly simple.

The best wisdom usually does.

You won't find much scolding in these pages.

The Earth has already endured enough shouting.

Instead, we'll take slow walks together.

We'll sit beside streams.

We'll examine mushrooms that appear overnight as if the forest had quietly changed its mind while we were sleeping.

We'll meet trees older than nations, insects smaller than commas, birds navigating by stars, and neighborhoods of life thriving beneath every footstep.

We'll wander through science, philosophy, history, psychology, and a few stories about a graybeard whose knees now provide weather forecasts with alarming consistency.

Along the way, I hope you'll discover something surprising.

The environmental movement isn't merely about saving forests.

It is about recovering relationships.

With places.

With seasons.

With neighbors both human and wild.

With gratitude.

With humility.

And perhaps, with the quieter, wiser version of yourself that still pauses to watch fireflies.

The planet, after all, has never really needed more owners.

It has always needed better borrowers.

Welcome.

Let's go for a walk.

Chapter One

The World Has Been Waiting Patiently

There is an old maple tree at the end of my street that has spent decades trying to get my attention.

For years, I failed the assignment.

I drove past it on my way to work.

I hurried beneath it carrying groceries.

I walked the dog under its branches while mentally rehearsing tomorrow's meeting, next week's project, and whether I had remembered to pay the electric bill. (For the record, I usually hadn't.)

The maple, displaying admirable patience, continued being a maple.

Every spring it unfolded thousands of new leaves with the confidence of an experienced stage actor.

Every summer it cast enough shade to cool the sidewalk for strangers it would never meet.

Every autumn it erupted into impossible shades of gold and crimson, as though the tree had quietly decided that if I wasn't going to notice subtlety, it would simply become spectacular.

Every winter it stood bare against gray skies, proving that dignity sometimes looks like endurance.

It never once complained about my inattention.

Trees have a remarkable capacity for emotional restraint.

Then one Tuesday morning, entirely by accident, I stopped.

Not because I had become enlightened.

Because I had dropped my car keys.

Some of the better moments in life begin with minor inconveniences.

While crawling around in the grass looking for a key ring that had somehow achieved military-grade camouflage, I noticed something I had never seen before.

The ground beneath that maple wasn't merely dirt.

It was busy.

Ants traveled invisible highways with astonishing efficiency.

Tiny mushrooms pushed through the soil after the previous night's rain.

A spider had suspended a web between two blades of grass so delicate it became visible only when the morning sun caught it.

A beetle, wearing what appeared to be polished black armor, walked with complete confidence despite being approximately the size of a raisin.

Above me, a chickadee landed on a branch, cocked its head as if evaluating my search technique, and flew away before offering any useful advice.

The maple itself stood quietly overhead, conducting an entire neighborhood of life that I had somehow mistaken for scenery.

I found my keys.

I also found a question.

How many extraordinary things had I been walking past every single day?

That question has followed me ever since.

Children ask questions like that naturally.

Adults schedule them for later.

One of the quiet tragedies of growing older is not that we lose intelligence.

It's that we often lose astonishment.

As children, we collected rocks because each one seemed different.

As adults, we mow around rocks because they complicate landscaping.

Children stop to watch caterpillars cross sidewalks.

Adults step over them while checking messages about productivity.

Children consider puddles invitations.

Adults consider them obstacles.

Somewhere between learning multiplication tables and learning how to reset internet routers, many of us unconsciously decided that familiarity was the same thing as understanding.

We began assuming that because we had seen something many times, we knew it.

We don't.

Not even close.

I've seen sunsets for decades.

I still have no idea how the sky manages to produce colors that painters spend lifetimes trying to imitate.

I've watched birds fly my entire life.

The engineering involved remains almost absurd.

I've looked at oak leaves every autumn without ever asking how a tree knows precisely when to let go.

Nature never became ordinary.

It simply became familiar enough that we stopped asking questions.

That distinction matters.

Imagine meeting someone at a neighborhood picnic.

You learn their name.

You discover they have three children, enjoy baking bread, and once backpacked across Iceland.

You would never say, "Well, I've met them once. I know everything there is to know."

Yet we make that assumption about the natural world all the time.

"That's just a squirrel."

Really?

A squirrel that remembers hundreds–sometimes thousands–of places where it buried food, helps plant future forests by forgetting some of those caches, leaps between branches with acrobatic precision, and survives winters that would send most of us searching for heavier sweaters before breakfast.

"That's just a robin."

A robin that navigates seasonal migrations using cues scientists are still unraveling, builds a nest sturdy enough to protect fragile eggs from spring storms, and spends dawn announcing its presence with a confidence many public speakers would envy.

"That's just a mushroom."

Except it isn't.

It is merely the visible portion of an immense underground network, a fruiting body briefly appearing above the soil while the larger organism quietly goes about the ancient work of recycling, connecting, nourishing, and sustaining the forest beneath our feet.

The closer science looks, the stranger and more beautiful the world becomes.

For centuries, we imagined forests as collections of individual trees competing for sunlight.

Then ecologists began discovering intricate underground partnerships linking roots and fungi in sprawling networks through which water, nutrients, and chemical signals move. Trees that appeared solitary turned out to be participants in relationships we had simply failed to notice.

The lesson reaches beyond forests.

Reality is often more connected than it first appears.

The same is true of rivers.

Of coral reefs.

Of prairies.

Of wetlands.

Of your own backyard.

Perhaps even of neighborhoods.

The great conservationist Aldo Leopold once suggested that we begin seeing land not as property but as a community to which we belong. It is a deceptively simple shift. Communities are not valued only for what they produce. They are valued because relationships have worth beyond utility.

You don't calculate the market value of a friendship every Tuesday morning.

You nurture it.

The living world deserves the same courtesy.

That may sound sentimental until science catches up.

Healthy forests filter drinking water.

Wetlands reduce flooding.

Bees pollinate crops.

Earthworms improve soil.

Mangroves soften storm surges.

Tiny plankton drifting through oceans produce much of the oxygen we breathe.

Relationships, it turns out, are practical.

Nature is not simply beautiful.

It is astonishingly competent.

Of course, competence often becomes invisible.

Think about plumbing.

Nobody thanks functioning plumbing until it stops functioning.

The same is true of ecosystems.

When birds sing, rivers flow, pollinators pollinate, fungi recycle nutrients, and forests regulate water and climate, we hardly notice.

Their success becomes background music.

Only silence gets our attention.

Rachel Carson understood this with heartbreaking clarity. She warned that a spring without birdsong would not merely be quieter. It would be telling us that countless invisible relationships had unraveled. Long before most people spoke about ecosystems, she reminded us that every action echoes outward in ways we seldom anticipate.

The older I become, the less interested I am in frightening people into caring about the Earth.

Fear has its place. It can awaken us to danger.

But love keeps us awake.

People drive hundreds of miles to visit mountains they have never seen.

They wake before sunrise to watch migrating cranes.

They kneel in gardens to coax tomatoes from the soil.

They carry children onto their shoulders so small eyes can see a hawk circling overhead.

None of this is required for survival.

It is required for belonging.

Perhaps that is what many of us have been missing.

Not information.

Relationship.

The Earth has never asked us to know everything.

It simply keeps inviting us to notice something.

A single leaf.

A bee dusted with pollen.

The smell of rain before the first drop falls.

The chorus of frogs on a warm evening.

The patient work of moss covering a stone.

Wonder rarely arrives wearing a spotlight.

It usually slips in through the side door disguised as an ordinary Tuesday.

When we learn to recognize it, something subtle changes inside us.

The tree at the end of the street stops being "that maple."

It becomes our maple.

The creek behind the neighborhood becomes our creek.

The hawk overhead becomes a familiar neighbor rather than anonymous wildlife.

The world gradually transforms from scenery into community.

And communities have a curious effect on us.

We care for them.

Not because someone ordered us to.

Because affection quietly rewrites our priorities.

That is why this book begins not with environmental problems but with attention.

Attention is where stewardship begins.

Before we restore rivers, we notice rivers.

Before we protect forests, we learn their names.

Before we worry about future generations, we rediscover the astonishing privilege of sharing this particular Tuesday with chickadees, maple trees, mushrooms, earthworms, clouds, and the countless neighbors whose lives intersect with ours in ways we are only beginning to understand.

The old maple at the end of my street is still there.

Every morning it goes about the dignified business of being exactly what it was created to be.

It asks nothing from me.

It sends no reminders.

It offers shade without invoices, beauty without advertising, oxygen without applause, and a quiet lesson without ever speaking a word.

I like to think it has forgiven me for all those years when I walked beneath its branches without really seeing it.

The remarkable thing about the living world is that it rarely keeps score.

It simply waits.

Patiently.

For us to look up.

Chapter Two

How We Accidentally Stopped Looking

Somewhere in my garage is a box containing enough extension cords to electrify a moderately sized carnival.

I have no explanation for this.

Like many men of a certain age, I apparently subscribe to the belief that every extension cord possesses unique emotional qualities that make it impossible to throw away.

There are orange ones.

Green ones.

One that no longer works but "might be useful for parts," though I couldn't tell you which parts.

I have never once looked inside that box and thought, You know what would make my life complete? Another extension cord.

And yet, somehow, they continue multiplying.

Attention works much the same way.

Every day we acquire thousands of little fragments of it.

News headlines.

Weather alerts.

Sales promotions.

Passwords.

Political arguments.

Sports scores.

Calendar reminders.

Text messages.

Notifications informing us that someone we vaguely remember from high school has posted a photograph of their lunch.

By evening, our minds resemble my extension cord box.

Everything is technically there.

Very little is organized.

And almost nothing is connected.

It wasn't always this way.

For most of human history, paying attention was a survival skill.

People noticed which berries were safe to eat.

They recognized the sound of approaching weather.

They could identify animal tracks in mud, distinguish healthy soil from exhausted soil, and predict seasonal changes by observing birds, insects, or the position of certain stars.

Observation wasn't a hobby.

It was breakfast.

Our ancestors lived in a world that constantly demanded awareness.

Today's world demands something quite different.

Reaction.

The average day now arrives with a parade of interruptions that would have bewildered previous generations.

A watch vibrates.

A phone chirps.

A laptop pings.

A television murmurs in the background while someone scrolls through another screen, all while a smart speaker politely offers tomorrow's forecast to a room that never asked.

The remarkable thing is not that technology exists.

The remarkable thing is how completely it occupies the tiny space between experience and awareness.

Have you ever reached the end of a walk and realized you remember almost nothing about it?

Your legs did the walking.

Your mind attended a meeting that wasn't there.

You traveled through a neighborhood without actually visiting it.

I've done this more times than I'd care to admit.

One afternoon, I walked nearly a mile while mentally composing an email I never sent.

The email consumed my attention.

The walk consumed my calories.

Neither accomplished very much.

When I finally looked up, I had no memory of the birds, the flowers, the changing clouds, or even the houses I had passed.

The world had patiently performed an entire symphony while I listened to the rehearsal in my own head.

That realization was humbling.

It also explained something I had noticed in myself.

As life became busier, the world seemed to become smaller.

Not physically.

Experientially.

Days blurred together.

Seasons arrived almost by surprise.

Spring became something I noticed because the lawn needed mowing.

Autumn announced itself through pumpkin-flavored products before I had even looked at the trees.

Winter appeared when my heating bill arrived.

Nature hadn't become less interesting.

I had become less available.

That distinction may be one of the most important in this book.

The problem is not that the world stopped speaking.

The problem is that we filled every quiet moment with something louder.

Silence has become one of the rarest habitats on Earth.

Not because birds stopped singing.

Because we started wearing earbuds.

The psychologist in me finds this fascinating.

Our brains evolved to notice novelty.

A sudden movement in tall grass.

An unfamiliar sound.

A flash of color.

These things deserved attention because they might represent opportunity–or danger.

Modern technology has become astonishingly skilled at manufacturing novelty.

Every notification whispers the same irresistible promise:

"Something new has happened."

Usually it hasn't.

Someone liked a photograph.

A package has moved three miles closer.

A headline predicts the collapse of civilization before lunch.

Our attention responds as though each alert were a rustle in ancient grass.

The biology is old.

The distractions are new.

This doesn't make technology the villain.

My goodness, no.

Technology has allowed me to see galaxies photographed by telescopes orbiting above the atmosphere.

It lets scientists monitor migrating whales, map disappearing glaciers, track endangered wildlife, and share discoveries that once took years to circulate.

It also lets me video chat with grandchildren who somehow know more about my television than I do.

Technology is a magnificent servant.

It simply makes a terrible landlord.

The question isn't whether we should use it.

The question is whether it quietly begins using us.

One of the great ironies of modern life is that we have more access to knowledge about nature than any generation in history.

We can identify birds with an app.

Recognize stars with another.

Learn the names of wildflowers by taking a photograph.

Watch bald eagles nesting live from hundreds of miles away.

Explore coral reefs without getting wet.

And yet many people know less about the living world immediately outside their own homes than their great-grandparents did.

We have become experts on distant places while remaining strangers to our own neighborhoods.

I once asked a classroom full of college students to name five native trees growing on campus.

Silence.

One young man confidently pointed toward a pine tree.

It was a flagpole.

In his defense, it was unusually straight.

Another student admitted she had never considered that trees had names beyond "tree."

Yet these same students could identify dozens of automobile models from remarkable distances.

Our attention had simply been trained elsewhere.

Attention is never accidental.

It always follows value.

We notice what we have learned is worth noticing.

That's why birdwatchers suddenly begin seeing birds everywhere.

Gardeners notice soil.

Astronomers notice the moon.

Musicians hear rhythms most people miss.

Parents become astonishingly skilled at recognizing the sound of one particular child crying across a crowded room.

The world hasn't changed.

The observer has.

Wonder works the same way.

It isn't something hidden from us.

It is something we gradually become capable of seeing.

Children possess this ability almost effortlessly.

Watch a young child discover a snail.

There is no hurry.

No agenda.

No productivity metric.

The snail receives complete attention.

Adults often smile at this as though children are being delightfully inefficient.

I have begun to suspect the children are onto something.

Perhaps efficiency has quietly convinced us to ignore the very experiences that make life meaningful.

A walk that produces no measurable outcome still nourishes the soul.

Watching fireflies creates no entry on a résumé.

Listening to rain contributes nothing to quarterly earnings.

Learning the names of neighborhood birds will not improve your tax return.

Yet these experiences enlarge us in ways difficult to quantify.

The natural world has always offered a different economy.

It rewards attention with belonging.

It pays curiosity in wonder.

It returns patience with understanding.

The dividends are invisible until one day you realize the world has become larger simply because you have learned to notice more of it.

That is why recovering our attention may be one of the most hopeful environmental acts available to us.

Before we can protect a forest, we have to see one.

Before we care about pollinators, we have to notice bees.

Before we restore streams, we have to hear water.

Conservation begins long before legislation.

It begins with awareness.

So here is a modest suggestion.

Tomorrow morning, leave your phone in your pocket for the first five minutes after you step outside.

Just five.

Look up.

Listen.

Notice which direction the wind is moving.

See whether the birds are already awake.

Find one plant you've never really looked at before.

Watch an ant long enough to become mildly impressed.

You needn't travel to a rainforest or climb a mountain.

The living world has never required exotic destinations.

It has always been quietly waiting outside your front door.

The Earth has not become less remarkable.

It is still performing miracles with embarrassing consistency.

Our task is simply to become available for them again.

Chapter Three

The Difference Between Seeing and Noticing

I have walked past the same crack in the sidewalk for nearly fifteen years.

I know this because one spring morning a tiny maple tree was growing out of it.

Not beside it.

Not near it.

Directly from the crack itself.

It couldn't have been more than four inches tall.

Two bright green leaves.

A stem no thicker than a toothpick.

The confidence of a creature that had absolutely no idea it was attempting the impossible.

I laughed out loud.

Not because it was funny.

Because it was wonderfully optimistic.

Somewhere the previous autumn, a maple seed had spun lazily through the air, landed in a sliver of accumulated dust, settled into a crack in the concrete, survived winter, found just enough moisture, just enough sunlight, and just enough determination to announce to the world:

"I believe this is an excellent place to become a tree."

It was, objectively speaking, a terrible place to become a tree.

The city eventually removed it.

Concrete almost always wins those arguments.

But for a few glorious weeks, that little maple reminded me of something I had been forgetting.

Life is astonishingly persistent.

It is constantly trying to happen.

Most of us simply fail to notice.

There is an important difference between seeing something and noticing it.

Seeing is biological.

Noticing is intentional.

Seeing happens because your eyes function.

Noticing happens because your mind and your heart decide something deserves attention.

We see thousands of things every day.

Road signs.

Coffee cups.

Door handles.

Traffic lights.

Most of them pass through our awareness like strangers crossing a train station.

Useful.

Necessary.

Forgotten moments later.

Noticing is different.

Noticing lingers.

It asks questions.

It invites curiosity.

It creates relationships.

That little maple wasn't just another plant.

It became a story.

Every morning on my walk, I looked for it.

Had it grown another leaf?

Had the rain helped?

Could it somehow survive?

A tiny volunteer tree had quietly become a neighbor.

The world changes the moment we begin assigning names instead of categories.

That isn't "a bird."

That's a chickadee.

That isn't "a flower."

That's a trillium.

That isn't "some tree."

That's a sugar maple.

Names slow us down.

They make anonymous things personal.

Think about the people in your life.

Imagine referring to your closest friend as "human."

Your spouse as "adult."

Your grandchild as "small person."

Technically accurate.

Emotionally ridiculous.

Relationships begin with names.

The natural world deserves the same courtesy.

The first time I learned to identify chickadees, I began hearing them everywhere.

Not because more chickadees had suddenly moved into the neighborhood.

Because my attention had finally been introduced.

Knowledge sharpens perception.

This is one of science's greatest gifts.

Science does not remove mystery.

It deepens it.

When I was young, I thought a spider web was simply sticky string arranged in attractive geometry.

Reasonably impressive.

Then I learned that spiders manufacture silk stronger than steel of the same weight, engineer different kinds of threads for different purposes, recycle portions of old webs by eating them, and somehow accomplish all of this without attending engineering school.

Now every spider web feels like an architectural conference.

The same happened with bees.

At first, they were simply insects that occasionally interrupted picnics.

Then I learned that many flowers have evolved shapes, colors, and even ultraviolet patterns specifically designed to guide pollinators toward nectar.

Bees, in turn, became participants in one of the oldest partnerships on Earth.

Suddenly a clover patch wasn't just grass with flowers.

It was an airport.

Every blossom a runway.

Every bee a tiny cargo pilot delivering life from one flower to another.

The world became richer because understanding made it more interesting, not less.

There is an old misconception that science replaces wonder.

The opposite is usually true.

Wonder often begins with ignorance.

It matures through understanding.

A rainbow is beautiful when you are six years old.

It remains beautiful after you learn about sunlight, water droplets, and the bending of light.

If anything, the explanation makes the phenomenon more astonishing.

Reality keeps turning out to be more imaginative than our first guesses.

That has happened to me repeatedly.

Trees turned out to communicate through fungal partnerships beneath the forest floor.

Octopuses solve puzzles.

Birds navigate across continents.

Whales sing songs that travel for miles through the ocean.

Tiny mosses survive conditions that would destroy far larger organisms.

The deeper we look, the less ordinary anything becomes.

Perhaps that is why curiosity has always been one of humanity's greatest survival tools.

Curiosity says,

"What is that?"

Wonder asks,

"Isn't that incredible?"

Wisdom quietly adds,

"What responsibility comes with knowing?"

Those three questions form a progression.

Curiosity opens the door.

Wonder invites us inside.

Wisdom asks us to behave like grateful guests.

The older I become, the more convinced I am that attention is one of the purest forms of respect.

When someone listens carefully while you speak, you feel valued.

When a teacher remembers your name, you feel seen.

When a friend notices you've been unusually quiet, you understand that someone cares.

Attention is how love often introduces itself.

Perhaps the same is true of the Earth.

The robin does not require your admiration.

The oak tree will continue growing whether you notice it or not.

The creek keeps flowing without applause.

Nature is wonderfully free from the need for compliments.

But we are changed by paying attention.

Every observation subtly rearranges us.

A child who spends an afternoon watching monarch butterflies is unlikely to feel indifferent toward milkweed.

A gardener who has nurtured tomatoes from tiny seeds rarely wastes food carelessly.

Someone who has watched a great blue heron standing motionless at dawn will forever smile when one flies overhead.

Relationship changes behavior.

That is why environmental stewardship begins long before recycling bins or national parks.

It begins with acquaintance.

You care about what has become familiar in the deepest sense–not merely recognized, but known.

I have a friend who knows nearly every bird call in our county.

Walking with him is like strolling through a neighborhood where everyone knows his name.

"There goes a wood thrush."

"Listen–that's a nuthatch."

"Hear the towhee scratching in the leaves?"

The woods sound entirely different when translated by affection.

I've also walked with botanists who stop every twenty feet to greet plants as though introducing old friends.

At first I thought this was slightly eccentric.

Then I realized they weren't slowing the walk.

They were enriching it.

One hour with someone who truly notices is often worth a dozen hurried hikes.

The poet in me envies this.

The scientist in me admires it.

The graybeard in me has finally begun practicing it.

These days I try to adopt what I call the "One New Thing Rule."

Whenever I go outside, I look for one thing I have never noticed before.

Just one.

A bird's nest hidden in the crook of a branch.

Lichen growing in delicate circles across a stone.

The fragrance of basswood blossoms.

The geometry inside a fern.

The remarkable way raindrops gather at the tips of pine needles before falling.

Finding one new thing each day doesn't sound like much.

Until you realize that by the end of a year, the world has become 365 wonders larger.

The remarkable thing is that the supply never runs out.

The Earth has been conducting this exhibition for billions of years.

Admission has always been free.

There is no gift shop.

No closing hour.

Only a standing invitation to pay closer attention.

The little maple tree in the sidewalk crack is long gone now.

Concrete reclaimed its territory.

But I still smile whenever I pass that spot.

Because I no longer see a crack.

I see a place where hope once tried to grow.

Perhaps that is what noticing does.

It teaches us that every ordinary place contains a story waiting patiently for someone willing to look long enough to discover it.

And once you begin discovering those stories, the world becomes wonderfully difficult to ignore.

Which is precisely where borrowing a planet begins.

Chapter Four
The Backyard Safari

I have been on safari exactly once.

It lasted twelve minutes.

The admission was free.

I forgot my binoculars.

The guide was a seven-year-old.

She lived three houses down.

One Saturday morning, she came over while I was pulling weeds from the flower bed. She stood at the edge of the driveway with the seriousness usually reserved for surgeons and airline pilots.

"Do you want to see something amazing?"

Now, at my age, this question deserves careful consideration.

Children generally define "amazing" differently than adults.

Amazing might mean a butterfly.

It might mean a worm wearing what appears to be a tiny mustache.

It might also mean watching them jump off the porch while wearing a towel tied around their neck.

Experience has taught me to ask one important question.

"Will I need to go to the emergency room?"

She looked puzzled.

"No."

"Excellent," I said. "Lead the way."

She marched confidently across the yard, around the maple tree, and stopped beside a patch of clover that I had intended to mow later that afternoon.

"There."

I looked.

"I don't see anything."

She sighed with the patience of someone explaining arithmetic to a particularly slow puppy.

"Not there."

She pointed six inches to the left.

"There."

I leaned closer.

Still nothing.

She folded her arms.

"You're too tall."

I had never considered this possibility.

I crouched.

Then knelt.

Finally, I lay on the grass, which caused my knees to file an immediate complaint with management.

And suddenly the world changed.

The clover patch became a city.

Honeybees drifted lazily from blossom to blossom, collecting nectar with quiet determination.

Tiny sweat bees, no larger than grains of rice, worked flowers I had not even noticed.

An ant navigated around pebbles that had become boulders.

A ladybug climbed the stem of a blade of grass with the confidence of an experienced mountaineer.

Hoverflies suspended themselves in midair with engineering that still seems slightly unfair.

Spiders had anchored invisible webs between stems.

A butterfly floated overhead like a brightly colored handkerchief caught by the breeze.

The little girl looked at me.

"See?"

I did.

Not because the insects had appeared.

Because I had finally entered their neighborhood.

For years I had treated my backyard as a piece of property.

It turns out it was an ecosystem.

Those are not the same thing.

Property belongs to us.

Ecosystems include us.

That distinction quietly changes everything.

When most people hear the word nature, they imagine spectacular places.

Towering mountains.

Ancient forests.

Thundering waterfalls.

National parks stretching beyond the horizon.

Those places deserve every bit of admiration they receive.

But they can accidentally teach us the wrong lesson.

They suggest that nature exists somewhere else.

Someplace we visit.

Someplace requiring hiking boots, expensive cameras, and enough granola bars to survive a minor expedition.

Nature, meanwhile, has been patiently occupying our neighborhoods.

It grows through sidewalk cracks.

It nests beneath porch roofs.

It blooms between fence posts.

It sings before dawn whether anyone is listening or not.

You don't have to travel to another continent to encounter the wild.

Sometimes you simply have to kneel down.

One of the greatest gifts science has given us is scale.

Astronomers taught us that we inhabit a very small planet in a very large universe.

Ecologists taught us that enormous worlds also exist in astonishingly small places.

Take a single teaspoon of healthy garden soil.

It looks unremarkable.

Brown.

Crumbly.

Perhaps slightly damp after a spring rain.

Yet inside that tiny spoonful lives an astonishing community.

Bacteria.

Fungi.

Protozoa.

Nematodes.

Microscopic insects.

Countless organisms interacting in ways scientists are still discovering.

The soil beneath your feet is less like dirt and more like a bustling city whose citizens have agreed to work the night shift.

Without them, forests fail.

Gardens fail.

Agriculture fails.

Eventually, so do we.

It is difficult to feel superior after learning that civilization depends upon creatures invisible to the naked eye.

Humility often arrives disguised as biology.

The same is true overhead.

One evening I stood outside with a neighbor who enjoys birds the way some people enjoy baseball statistics.

He pointed toward a dead tree.

"What do you see?"

"A dead tree."

He smiled.

"What else?"

I looked harder.

Nothing.

Then he handed me his binoculars.

Woodpecker holes.

Bluebirds.

A squirrel.

A tiny vine climbing the trunk.

Fungi spreading beneath loose bark.

Insects moving in and out of narrow crevices.

The dead tree wasn't dead at all.

It had simply changed occupations.

It had become an apartment building.

The forest wastes very little.

One creature's ending quietly becomes another's beginning.

Nature has always been an excellent recycler.

Long before humans invented the phrase "circular economy," fallen leaves were becoming soil, fallen trees were becoming homes, and mushrooms were performing the patient work of turning yesterday into tomorrow.

Nothing lectures.

Nothing advertises.

The process simply continues.

Perhaps that is why the natural world is such an extraordinary teacher.

It rarely explains itself.

It demonstrates.

A tree never gives a speech about generosity.

It simply offers shade.

A flower never campaigns for pollinators.

It opens.

A river never announces persistence.

It keeps flowing.

The lesson is always embodied before it is spoken.

One of my favorite naturalists once wrote that if you spend an entire year observing one small patch of forest, you will witness enough change to occupy a lifetime of curiosity.

I believe him.

In fact, I think the same could be said for an ordinary backyard.

Spring begins with birds inspecting nesting sites while trees quietly inflate buds that seemed lifeless only weeks before.

Summer fills every corner with astonishing industry.

Bees work.

Spiders weave.

Dragonflies patrol.

Young birds practice flying with all the grace of teenagers learning to parallel park.

Autumn becomes a festival of preparation.

Seeds scatter.

Squirrels hide acorns with theatrical urgency.

Goldenrod feeds late pollinators.

Leaves drift downward, beginning the slow work of becoming next year's soil.

Winter appears quiet until you slow down.

Rabbit tracks.

Cardinals bright against fresh snow.

The geometry of bare branches.

Moss remaining green despite the cold.

Even apparent stillness turns out to be full of waiting.

The backyard never repeats itself.

It only appears to.

The older I get, the more convinced I become that boredom often says less about the world than it does about the observer.

Children understand this instinctively.

Give them an empty cardboard box and they discover a spaceship.

Give them a stick and they invent a sword, a fishing pole, a magic wand, and an archaeological tool before lunch.

Adults often reverse the equation.

We stand in the middle of astonishing abundance and declare there is nothing interesting to see.

Perhaps the problem is not imagination.

Perhaps it is altitude.

We spend too much of life standing up.

Standing is excellent for reaching high shelves.

It is less effective for discovering beetles.

The Earth looks remarkably different from three feet lower.

That little girl taught me that.

She didn't know the Latin names of insects.

She couldn't explain pollination or food webs or ecosystem services.

She simply knew that paying attention was fun.

Scientists spend careers developing sophisticated ways to observe nature more carefully.

Children begin there.

Maybe wisdom is not becoming more complicated.

Maybe wisdom is recovering what curiosity knew all along.

Since that Saturday morning, I have developed a habit.

Whenever I enter a new place–a park, a trail, even someone's backyard–I pause for one minute before walking.

I don't do anything.

I simply let the place introduce itself.

The wind usually speaks first.

Then the birds.

Then the insects.

Gradually details emerge.

A hawk circling overhead.

Milkweed beside the fence.

Lichens painting old stones.

The faint fragrance of pine.

The place was always saying hello.

I had merely arrived talking.

The world rewards those who arrive listening.

You need not become an ecologist to borrow this planet well.

You don't have to memorize every bird call or identify every wildflower.

You don't have to own expensive binoculars or hiking boots or field guides thick enough to qualify as weightlifting equipment.

You simply have to become willing to notice your own neighborhood with the curiosity of someone seeing it for the first time.

Because the greatest safari most of us will ever take is not across an African savanna or through an Amazon rainforest.

It is across the twenty-five feet between the back door and the old maple tree.

The distance is short.

The journey, it turns out, is immense.

Chapter Five

The Last Time You Looked at a Cloud

I have reached the age where I occasionally walk into a room and forget why I went there.

This happens often enough that I have developed a sophisticated retrieval system.

I simply stand there until the room reminds me.

Sometimes it works.

Sometimes I leave with a flashlight, a screwdriver, and a bowl of peanuts without ever remembering the original mission.

Growing older is a fascinating experiment.

One afternoon, while standing in the backyard trying to remember why I had gone outside in the first place, I looked up.

Not because I was being philosophical.

Because I had become temporarily directionless.

Above me drifted enormous white clouds against an impossibly blue sky.

I don't know how long I stood there.

Long enough for one cloud that resembled a sailing ship to become something that looked remarkably like a sleeping golden retriever.

Clouds, unlike most politicians, are perfectly comfortable changing their minds.

Eventually I remembered why I had gone outside.

It no longer seemed very important.

I had accidentally spent ten quiet minutes doing something I had not intentionally done in years.

I had watched clouds.

Children do this naturally.

Adults generally require one of three conditions.

A beach.

A power outage.

Or a doctor telling them they need to reduce stress.

Somewhere along the journey into adulthood, we quietly classify cloud watching as unproductive.

That's unfortunate.

Clouds have been trying to teach us for a very long time.

The first lesson is that nothing stands still.

Not rivers.

Not forests.

Not oceans.

Not weather.

Not us.

Nature is never frozen in time.

It is movement pretending to be permanence.

The oak tree at the end of your street looks the same year after year.

It isn't.

Every day it converts sunlight into energy.

It grows new wood.

Extends tiny roots.

Feeds fungi beneath the soil.

Repairs damaged bark.

Produces leaves.

Drops leaves.

Shelters birds.

Hosts insects.

Recycles nutrients.

Breathes.

Lives.

The tree appears still only because our lives move at a different speed.

Mountains teach the same lesson.

We imagine mountains as symbols of permanence.

Then geologists quietly explain that mountains rise.

Mountains erode.

Mountains travel.

Entire continents drift across the face of the Earth so slowly that only deep time reveals the journey.

The world is not a collection of objects.

It is a collection of processes.

That realization changes how we see almost everything.

A river is not water.

It is flowing.

A forest is not trees.

It is becoming.

A meadow is not grass.

It is countless relationships unfolding simultaneously.

Even a cloud isn't really a thing.

It is a moment.

A brief gathering of water vapor, sunlight, wind, temperature, and atmospheric conditions that has agreed to exist together for a little while before becoming rain, mist, snow, or memory.

Perhaps that is why clouds have always fascinated poets.

They understand impermanence.

Science simply explains how it works.

One of the great gifts of growing older is discovering that science and poetry are often describing the same reality from different directions.

The poet says the forest breathes.

The scientist explains photosynthesis.

Both are pointing toward wonder.

The poet says rivers remember.

The hydrologist describes watersheds, sediments, and seasonal flow.

Again, both are trying to describe something larger than themselves.

Reality is generous enough to satisfy both.

There is another lesson hidden in the sky.

Perspective.

Have you ever noticed how dramatically the world changes simply by looking upward?

Our lives naturally pull our attention toward the ground.

Bills.

Schedules.

Deadlines.

Laundry.

Traffic.

The grocery list you forgot to bring into the store.

Gravity affects more than our bodies.

It pulls at our attention.

Looking upward interrupts that pull.

Birds remind us.

Clouds remind us.

Stars remind us.

There is always more above us than our worries.

The older I become, the more I suspect that many of humanity's greatest discoveries began because someone looked where everyone else wasn't looking.

Astronomers looked up.

Geologists looked down.

Naturalists looked closely.

Explorers looked beyond the horizon.

Children looked under rocks.

The pattern is remarkably consistent.

Progress often begins with changing the direction of our attention.

I remember lying in the grass as a boy, watching clouds drift across a summer sky while my grandfather stretched out beside me.

He wore an old straw hat that had survived more seasons than I had birthdays.

"What do you see?" he asked.

"A dragon."

He nodded.

"I see an old fisherman."

A few minutes later neither of us was correct.

The dragon had become a rabbit.

The fisherman looked suspiciously like Nebraska.

Neither of us was disappointed.

The point had never been accuracy.

The point was paying attention together.

I didn't appreciate it then, but my grandfather was teaching me a way of seeing.

He wasn't filling my head with information.

He was making room for observation.

Those are not the same thing.

Information answers questions.

Observation creates them.

The environmental movement has given us remarkable amounts of information.

We know more about ecosystems, biodiversity, climate, soil, oceans, forests, and wildlife than any previous generation.

Knowledge matters.

It saves lives.

It guides decisions.

It shapes policy.

But information alone rarely changes hearts.

Observation does.

No graph has ever replaced the experience of standing beneath a sky filled with migrating geese.

No statistic has ever fully captured the feeling of hearing spring peepers after a long winter.

No chart can duplicate the first time a child discovers a monarch butterfly emerging from its chrysalis.

Experience gives information a home.

Perhaps that is why so many people who devote their lives to conservation have stories rather than arguments.

Ask them why they care.

They'll tell you about a river.

A forest.

A grandmother's garden.

A hawk.

A camping trip.

A tide pool.

A favorite fishing spot.

A night beneath astonishing stars.

Love usually arrives before advocacy.

Memory often precedes commitment.

The Earth rarely changes us all at once.

It works slowly.

One sunrise.

One snowfall.

One thunderstorm.

One unexpected rainbow.

One quiet evening watching clouds drift across an endless sky.

I sometimes wonder whether our ancestors understood this better than we do.

Before clocks divided the day into appointments, people read the sky.

Clouds announced weather.

Winds carried news.

The moon measured time.

Stars guided travelers home.

The heavens were not scenery.

They were conversation.

Today, satellites can predict tomorrow's weather with astonishing accuracy.

I am deeply grateful for that.

I also wonder whether something was lost when we stopped looking up for ourselves.

Convenience occasionally comes with invisible costs.

One of those costs may be attention.

Fortunately, attention is remarkably recoverable.

It returns the moment we choose to offer it.

The next time you find yourself outside with nowhere urgent to be, look up.

Not because you expect something dramatic.

Because the sky has been faithfully putting on a new performance every single day of your life.

No two sunsets have ever been identical.

No cloud has ever followed precisely the same path.

No storm has repeated itself exactly.

Creation appears to have a delightful dislike for copies.

The sky refuses to become ordinary.

Perhaps that is why it continues inviting us to become less ordinary observers.

Eventually, those few minutes spent watching clouds begin to change something deeper.

You become more patient.

More attentive.

More willing to let moments unfold instead of rushing them toward conclusions.

You begin noticing the weather before the forecast.

You recognize the first hints of autumn in the angle of the sunlight.

You smell rain before it arrives.

You hear thunder farther away than you once could.

The world has not become more communicative.

You have simply become a better listener.

Borrowing a planet well is not only about caring for forests and rivers.

It is about recovering our membership in a conversation that has been taking place for billions of years.

The trees are speaking.

The birds are speaking.

The soil is speaking.

Even the clouds, drifting quietly overhead, have something to say.

The question is no longer whether the Earth is communicating.

The question is whether we have remembered to look up long enough to listen.

Chapter Six

The Humbling Mathematics of Trees

There is an old white oak in a nearby park that has quietly become one of my professors.

It has never assigned homework.

It has never charged tuition.

It has never once corrected my grammar, which is generous considering some of the sentences I've inflicted upon the English language over the years.

It simply stands there.

Every season.

Every year.

Patiently conducting a class that most people hurry past.

The first time I really paid attention to it, I did what educated people often do when confronted with something magnificent.

I looked up the statistics.

Approximately ninety feet tall.

Canopy spreading nearly as wide.

Hundreds of years old.

Thousands upon thousands of leaves.

Tons of wood.

Millions of acorns over its lifetime.

Interesting numbers.

Impressive numbers.

Useful numbers.

They also missed the point.

Numbers can describe the size of a tree.

They cannot describe the experience of standing beneath one.

One afternoon I leaned against its trunk and tried to imagine everything that tree had witnessed.

When it was a young sapling, Abraham Lincoln had not yet been born.

Generations of children had climbed its lower branches.

Couples had carved initials into its bark–an unfortunate tradition that trees have tolerated with considerably more grace than I would.

Storms had broken limbs.

Lightning may have scarred it.

Droughts had tested it.

Winters had stripped it bare.

Spring had restored it.

Again.

And again.

And again.

The tree had quietly survived events that entire civilizations now remember only in history books.

Meanwhile, I become impatient if the microwave has eighteen seconds remaining.

Perspective is an excellent treatment for self-importance.

Trees administer it freely.

We humans have an interesting relationship with time.

A five-minute wait at the doctor's office feels excessive.

A two-hour meeting feels eternal.

A thirty-year mortgage seems unimaginably long.

Trees smile politely.

Some bristlecone pines have been alive since before the construction of the pyramids.

Giant sequoias were already old when Julius Caesar was young.

Ancient olive trees have watched empires appear and disappear with something approaching botanical indifference.

Compared to them, we are enthusiastic mayflies.

This is not depressing.

It is wonderfully liberating.

We spend enormous amounts of energy trying to become important.

Trees seem content becoming useful.

An oak does not wake each morning wondering whether it has achieved enough.

It simply continues being an oak.

Providing shade.

Holding soil.

Sheltering birds.

Feeding insects.

Filtering air.

Storing carbon.

Quietly making the world more livable.

No press releases.

No motivational seminars.

No personal branding strategy.

Just faithful participation.

I confess that I admire this.

There comes an age when applause begins losing some of its charm.

Contribution becomes far more interesting.

Science has only increased my admiration.

When I was young, I thought trees were mostly made of wood.

Technically correct.

Magnificently incomplete.

Trees are also water towers.

Air conditioners.

Apartment buildings.

Restaurants.

Nurseries.

Chemical laboratories.

Weather moderators.

Conversation partners with fungi.

Solar-powered manufacturing plants.

Neighborhood gathering places.

A mature oak can support hundreds of species.

Birds nest in its branches.

Caterpillars feed on its leaves.

Woodpeckers search its bark for insects.

Squirrels cache its acorns.

Fungi partner with its roots.

Moss settles into its crevices.

Lichens paint quiet masterpieces across aging bark.

Even after the tree dies, it continues serving others for decades.

One organism.

Hundreds of relationships.

It is difficult to remain arrogant after contemplating that.

One of my favorite discoveries about trees came from an ecologist who remarked that forests are less like collections of individuals and more like communities.

At first, that seemed oddly sentimental.

Then the science arrived.

Roots connect with fungal networks.

Nutrients move.

Chemical warnings spread.

Seedlings often receive support from older trees during their most vulnerable years.

The forest behaves less like a competitive marketplace and more like a neighborhood.

Now, before we become too romantic, trees are not holding town hall meetings.

They are not voting on zoning ordinances.

But they are participating in relationships far richer than we once imagined.

Perhaps the greatest scientific discoveries don't make the world less mysterious.

They simply reveal that we have been asking questions that were too small.

The mathematics of a tree are astonishing.

Consider a single leaf.

It quietly captures sunlight.

Draws water upward from roots that may extend farther than the tree is tall.

Exchanges gases with the atmosphere.

Manufactures sugars from air, water, and light through the elegant chemistry of photosynthesis.

Then, at the end of the season, gracefully lets go.

Multiply that by hundreds of thousands of leaves.

Then multiply again by millions of trees.

The numbers become almost absurd.

Forests are among the largest solar-powered systems on Earth.

They accomplish this without electrical outlets.

Nature has always had a flair for elegant engineering.

Yet the greatest lesson trees offer may have nothing to do with biology.

It has to do with scale.

Most of us measure our lives in years.

Trees invite us to think in generations.

The shade I enjoy today often comes from someone who planted a tree knowing they would never experience its full grandeur.

That strikes me as one of the most hopeful acts a human being can perform.

To invest in a future you will never personally inhabit.

To improve a world you will eventually leave.

To plant confidence instead of certainty.

I have often wondered what was going through the mind of the person who planted that old white oak.

Perhaps nothing profound.

Maybe they simply wanted shade.

Maybe they liked acorns.

Maybe they had no idea they were creating a place where a

graybeard would someday stand with a notebook, trying to make sense of life.

We rarely know the full consequences of our kindness.

Fortunately, goodness does not require complete information.

There is another mathematical truth trees quietly teach.

Growth is usually invisible.

Walk outside tomorrow morning and stare at a tree.

Nothing appears to happen.

Return the following day.

Still nothing.

The next week.

The next month.

The next season.

The tree seems almost unchanged.

Then return ten years later.

Transformation.

Growth rarely announces itself while it is happening.

The same is true of character.

Friendship.

Communities.

Children.

Forests.

Even civilizations.

The greatest changes often occur beneath the threshold of daily perception.

That realization has made me more patient with both trees and people.

Not everything meaningful grows quickly.

The older I become, the less impressed I am by things that appear overnight.

Old forests.

Strong marriages.

Wise teachers.

Trustworthy friends.

Healthy communities.

These are all slow achievements.

The world keeps trying to convince us that faster is better.

Trees politely disagree.

They suggest another possibility.

Deeper is better.

One ring at a time.

One season at a time.

One generation at a time.

Perhaps that is why forests have always felt like cathedrals.

Not because they are silent.

Forests are gloriously noisy.

Leaves whisper.

Woodpeckers hammer.

Wrens argue.

Wind sighs through branches.

Insects hum.

Streams chatter nearby.

A cathedral is not defined by silence.

It is defined by perspective.

You enter feeling large.

You leave feeling appropriately small.

That is one of the finest gifts the natural world offers.

Not the humiliation of insignificance.

The freedom of belonging to something immeasurably larger than yourself.

The old white oak has taught me many things.

That patience often outlives urgency.

That usefulness is more enduring than fame.

That quiet faithfulness changes the world more reliably than dramatic gestures.

That communities are stronger than individuals standing alone.

And perhaps most importantly, that a well-lived life is not measured by how much space we occupy, but by how much life flourishes because we were here.

Trees never seem worried about whether anyone notices them.

They simply keep growing toward the light.

There is wisdom in that.

More than most of us manage before our second cup of coffee.

Chapter 7

The Secret Social Lives of Forests

But something remarkable is happening.

Chemical signals move.

Resources move.

Warnings move.

When insects attack one part of a forest, neighboring trees sometimes begin producing defensive chemicals before the insects arrive.

Seedlings growing in deep shade may receive support through these underground partnerships during their earliest years.

The forest behaves less like isolated individuals competing for survival and more like a community whose members are deeply interconnected.

That idea changed something in me.

Not because it made forests sentimental.

Because it made them familiar.

Healthy human communities work much the same way.

The strongest neighborhoods are not those with the tallest fences.

They are the ones where people quietly help one another.

Someone shovels an elderly neighbor's sidewalk.

Someone drops off soup after surgery.

Someone keeps an eye on the house while a family is away.

No speeches.

No headlines.

Simply countless small acts of cooperation that make the whole neighborhood stronger.

Forests appear to have reached similar conclusions millions of years before we did.

Nature has always been a patient teacher.

One lesson keeps appearing in different forms.

Life succeeds through relationship.

Our culture often celebrates rugged independence.

We admire the self-made individual.

The lone genius.

The solitary hero.

Nature smiles kindly.

Nothing is truly self-made.

The oak depends upon fungi.

The fungi depend upon sugars.

The bird depends upon insects.

The insects depend upon flowers.

The flowers depend upon pollinators.

The pollinators depend upon healthy habitat.

Even the healthy soil depends upon countless microscopic organisms carrying out invisible work beneath our feet.

The closer we look, the less alone anything becomes.

That realization can be wonderfully humbling.

Years ago, I visited an old-growth forest where many of the trees were older than my country's founding.

Standing there, I felt what I suspect many people feel in places like that.

Very small.

Very grateful.

Very quiet.

I noticed enormous fallen logs covered in moss.

My first thought was predictable.

"What a shame."

Then the ranger smiled.

"Those are some of the busiest places in the forest."

He was right.

Closer inspection revealed mushrooms emerging from decaying wood.

Young trees rooting atop ancient trunks.

Salamanders hiding beneath loose bark.

Insects tunneling through softened wood.

Ferns carpeting the surface.

What looked like death from a distance had become astonishing generosity up close.

The fallen tree hadn't stopped contributing.

It had simply changed its job.

I find that deeply comforting.

Perhaps usefulness survives many endings.

There is another lesson hidden beneath the forest floor.

Most important work happens where almost nobody is looking.

The fungal networks connecting trees remain largely invisible.

Healthy roots stay underground.

So does the patient labor of earthworms.

We celebrate flowers.

We photograph autumn leaves.

Meanwhile, the unseen workers continue making those visible beauties possible.

Human communities operate similarly.

Teachers.

Caregivers.

Volunteers.

Custodians.

Neighbors quietly checking on neighbors.

Parents reading bedtime stories.

Most of civilization rests upon people whose names rarely appear in newspapers.

Forests understand this perfectly.

The visible canopy receives admiration.

The invisible partnerships sustain everything.

Perhaps our own lives would become healthier if we admired invisible faithfulness as much as visible achievement.

The older I become, the more impressed I am by people who quietly make other lives better.

A woman who tends a community garden.

A man who plants trees whose shade he will never enjoy.

The volunteer who removes litter from a creek without posting photographs afterward.

The grandmother who teaches children the names of birds.

These people resemble fungi more than celebrities.

I assure you that is a compliment.

One of the remarkable characteristics of healthy ecosystems is resilience.

Storms arrive.

Branches fall.

Lightning strikes.

Drought comes.

Insects appear.

Some trees die.

Yet forests recover.

Not because individual trees are indestructible.

Because relationships remain.

The network survives.

Human beings often misunderstand resilience.

We imagine it means never breaking.

Nature suggests something wiser.

Resilience means remaining connected.

When one part suffers, another helps.

When conditions improve, growth resumes.

The system adapts.

That is true of forests.

It is equally true of families, neighborhoods, friendships, and nations.

Perhaps this explains why loneliness has become one of the great hidden environmental issues of our time.

Not because lonely people damage forests.

Because loneliness separates us from the habits that make stewardship possible.

People who feel connected are more likely to care for places.

People who know their neighbors tend to notice when streams become polluted, when trees disappear, or when birds stop singing.

Connection encourages responsibility.

Isolation encourages indifference.

The forest has been quietly demonstrating this for millions of years.

Nobody thrives alone.

There is a phrase I have come to treasure.

"Everything is connected."

We hear it so often that it risks becoming wallpaper.

A pleasant slogan.

An environmental bumper sticker.

Science keeps insisting it is literal.

Pull too hard on one thread and another moves.

Protect one species and dozens of others benefit.

Destroy one habitat and consequences ripple outward.

Restore one wetland and entire watersheds become healthier.

Plant native flowers and butterflies return.

Healthy systems rarely improve one piece at a time.

They improve through relationships.

That may also be true of people.

The greatest moments in my own life have almost never happened in isolation.

Someone taught me.

Someone encouraged me.

Someone forgave me.

Someone believed in me before I deserved it.

Looking back, I realize I have spent my life being supported by networks I could scarcely see.

Teachers.

Family.

Friends.

Colleagues.

Authors I never met.

Scientists whose discoveries quietly changed how I understand the world.

Perhaps we are all standing because unseen roots have been holding us up.

There is something profoundly hopeful about that.

The forest reminds us that strength is not merely the ability to stand alone.

Sometimes strength is the willingness to remain connected.

When I walk through the woods now, I no longer see rows of separate trees.

I see conversations.

Not conversations carried by words.

Conversations carried by generosity.

By exchange.

By mutual dependence.

By the quiet agreement that thriving together is wiser than surviving alone.

It occurs to me that this is also an excellent way to borrow a planet.

Not as isolated consumers occupying the same address.

But as neighbors participating in the same living community.

The forest has understood that from the beginning.

It has simply been waiting for us to catch up.

Chapter Eight

The Mushroom Is Doing More Than You Think

I used to believe mushrooms had three basic purposes.

The first was to appear on pizza.

The second was to appear in soup.

The third was to surprise me by appearing in the lawn the day after a heavy rain.

This, I now realize, was an embarrassingly limited education.

Like many people, I spent most of my life noticing mushrooms only when they interrupted my plans.

One had appeared in the flower bed.

Another had emerged between the stepping stones.

A tidy circle of them had mysteriously formed in the backyard overnight, prompting neighborhood speculation ranging from fairy tales to unusually ambitious squirrels.

Then I met a mycologist.

Every field of science has people whose enthusiasm is so contagious that you begin caring deeply about subjects you didn't know existed an hour earlier.

Bird people have this effect.

Astronomers certainly do.

Mycologists are in a category entirely their own.

Show them a mushroom the size of your thumb, and they react the way sports fans react to a championship-winning goal.

They kneel.

They grin.

They begin using words like mycelium with such affection that you find yourself nodding along, hoping nobody asks you to spell it.

What they see is not a mushroom.

They see the brief appearance of something much larger.

The mushroom itself is only the fruit.

The real organism is usually hidden underground.

Imagine walking through an apple orchard and believing the apples were the tree.

That was essentially my understanding of fungi.

The important part remained invisible.

This seems to be one of nature's favorite jokes.

Again and again, the most essential work happens where we cannot easily see it.

Roots.

Soil microbes.

Groundwater.

Seeds.

The slow formation of fertile earth.

Even character works this way.

We spend years admiring visible success while quietly overlooking the invisible habits that made it possible.

Nature keeps repeating the lesson.

Pay attention to what is beneath the surface.

The underground network of fungal threads, called mycelium, stretches through soil with astonishing reach.

These delicate strands gather water.

Transport nutrients.

Break down fallen leaves and dead wood.

Recycle life's leftovers into tomorrow's beginnings.

Without fungi, forests would eventually bury themselves beneath generations of undecayed leaves and fallen trees.

The forest would become a magnificent warehouse with no custodial staff.

Fortunately, fungi arrived millions of years before we did and accepted the position without requesting benefits.

They perform one of the least glamorous jobs in creation.

Cleaning up.

There are not many awards for decomposition.

Nobody gathers around a fallen log to applaud the patient work of decay.

Children do not announce that they want to grow up and become world-famous recyclers of organic matter.

Yet without decomposition, life would slowly grind to a halt.

Dead things would remain dead.

Nutrients would remain trapped.

New growth would struggle to begin.

The Earth depends on creatures remarkably comfortable with endings.

There is wisdom in that.

Human beings often divide the world into exciting work and ordinary work.

Nature doesn't seem to recognize the distinction.

Every task matters.

The bee pollinating flowers.

The earthworm loosening soil.

The mushroom dissolving yesterday's fallen branch.

The bacteria recycling nutrients.

No occupation is glamorous.

Every occupation is essential.

The older I become, the more I admire work that quietly allows other life to flourish.

Perhaps that's because I've known people like that.

Teachers who stayed after school to encourage struggling students.

Custodians who somehow made old buildings feel welcoming.

Neighbors who shoveled sidewalks before anyone asked.

Volunteers who stacked chairs after everyone else had gone home.

They rarely attracted much attention.

They made everything else possible.

Forests are full of such people.

They simply happen to have spores instead of driver's licenses.

One autumn afternoon I found myself hiking after several days of rain.

The forest floor looked as though someone had scattered tiny umbrellas everywhere.

Orange mushrooms.

White mushrooms.

Purple mushrooms.

Mushrooms resembling tiny cups.

Others shaped like coral.

Some looked almost elegant.

Others appeared to have been designed by a committee that had entirely given up on symmetry.

Nature has a delightful refusal to be boring.

I knelt beside one no larger than a coin.

It had emerged from a decaying stick.

A year earlier, I would have walked right past it.

Now I found myself wondering.

How long had the hidden network been growing beneath the soil?

What invisible work had preceded this tiny appearance?

What partnerships had made this possible?

Curiosity has a wonderful way of replacing indifference.

Science has made fungi even stranger.

Some species help trees gather nutrients from soils they could never reach alone.

Others protect roots from disease.

Still others help plants survive drought.

Certain fungi communicate chemically with surrounding organisms in ways researchers are only beginning to understand.

The more scientists investigate, the more they discover that forests are built not only from trunks and branches but also from astonishing partnerships hidden beneath every footstep.

It reminds me of something that happened years ago while teaching.

A student visited during office hours.

She wasn't struggling academically.

She simply wanted to thank me for something I had completely forgotten.

Months earlier, after class, I had casually encouraged her to apply for an internship she believed was beyond her abilities.

She applied.

She was accepted.

That internship eventually shaped her career.

I had no memory of the conversation.

For her, it changed everything.

The experience has stayed with me because it revealed how often invisible acts produce visible consequences years later.

Nature understands this perfectly.

The fungus does not receive applause when the forest thrives.

The roots receive little recognition when branches reach sunlight.

Healthy soil rarely appears on postcards.

Invisible faithfulness quietly supports visible beauty.

Perhaps this is why humility feels so at home in the natural world.

Nothing insists on being the center of attention.

The flower depends on the bee.

The bee depends on the flower.

The tree depends on the fungus.

The fungus depends on the tree.

The forest depends on countless relationships, each contributing something unique while claiming remarkably little credit.

Humility, I have come to suspect, is not thinking less of yourself.

It is understanding that you are part of something much larger than yourself.

Mushrooms understand this.

Or perhaps it would be more accurate to say they simply live it.

Another lesson emerged as I learned more about fungi.

Waste is often a failure of imagination.

In forests, yesterday's leaves become tomorrow's soil.

Dead trees become nurseries.

Fallen branches become homes.

Nothing is merely discarded.

Everything becomes available for new purpose.

Human beings have not always adopted this philosophy.

We excel at producing things designed to be thrown away.

Nature appears genuinely puzzled by the concept.

It asks a wonderfully practical question.

"What else could this become?"

That may be one of the most hopeful questions we can ask about almost anything.

A broken neighborhood.

A struggling student.

A retired teacher.

A vacant lot.

A forgotten stream.

A fallen tree.

A mistake.

A failure.

What else might this become?

The forest answers every day.

One life feeds another.

One ending prepares a beginning.

One generation nourishes the next.

Nothing meaningful is ever truly isolated.

The cycle continues.

When I was younger, I tended to admire things that were impressive.

Tall mountains.

Large waterfalls.

Ancient redwoods.

Those still deserve admiration.

But now I also admire the things that make those wonders possible.

The unseen.

The overlooked.

The patient.

The persistent.

The mushroom pushing quietly through damp leaves after a night of rain.

The invisible network beneath it.

The slow transformation of decay into fertility.

The remarkable confidence with which life refuses to waste anything.

Including our failures.

Including our endings.

Including us.

Perhaps borrowing a planet well means learning to value not only what is beautiful, but also what quietly makes beauty possible.

That realization has changed the way I walk through forests.

I still notice towering trees.

I still pause beside waterfalls.

I still look upward.

But every now and then I stop for something no larger than my thumb.

A humble mushroom emerging from yesterday's fallen branch.

A reminder that some of the greatest work on Earth happens out of sight, asks for no applause, and leaves the world immeasurably richer than it found it.

There are worse ways to spend a life.

]

Chapter Nine

Everything Eats. Everything Is Eaten.

One of the more awkward moments in my teaching career occurred when a student asked me whether nature was fundamentally kind.

Professors are expected to answer questions.

Graybeards are expected to pause first.

I did.

Because the honest answer is more complicated than either of us wanted.

Nature is astonishingly generous.

Nature is also astonishingly honest.

Those are not always the same thing.

The robin feeding her chicks does not apologize to the caterpillar.

The fox does not conduct an ethics seminar before chasing the rabbit.

The owl does not submit a written explanation to the mouse.

The oak tree drops thousands of acorns, fully aware that many will become meals instead of forests.

Life feeds life.

It always has.

The first time many children encounter this reality is memorable.

Perhaps they're watching a hawk capture a squirrel.

A spider wrapping a fly.

A praying mantis making decisions that would earn disapproving looks at a family dinner.

Their instinct is often to intervene.

"Save it!"

It is one of the sweetest impulses we possess.

We want the small creature to live.

We want every story to have a happy ending.

The difficulty is that in nature, every happy ending for one creature often depends upon a difficult ending for another.

This isn't cruelty.

It is ecology.

The forest is not a courtroom.

It is a community where energy keeps changing hands.

Every living thing requires nourishment.

Plants gather sunlight.

Herbivores eat plants.

Carnivores eat herbivores.

Decomposers eventually recycle everyone.

The cycle is neither sentimental nor malicious.

It simply is.

That realization can be unsettling.

It can also be strangely comforting.

Because it reminds us that life is built not upon domination but upon participation.

Everything receives.

Everything gives.

Eventually.

When I was younger, I imagined the food chain as a ladder.

One species above another.

The mighty at the top.

The weak below.

Science has steadily replaced that image with something far more interesting.

The natural world is less like a ladder and more like a web.

Remove one strand, and others begin to shift.

Eliminate wolves from an ecosystem, and deer populations may expand beyond what forests can sustain.

The forests change.

Streams change.

Bird populations change.

Even insects may change.

Reintroduce the wolves, and astonishing recoveries sometimes ripple outward through the entire landscape.

Ecologists call these trophic cascades.

I call them reminders that reality is rarely simple.

Everything matters.

Even the things we overlook.

Especially the things we overlook.

One of my favorite examples involves sea otters.

Sea otters eat sea urchins.

Sea urchins eat kelp.

When otter populations decline, urchins multiply.

The expanding urchin population consumes vast underwater kelp forests.

Fish lose habitat.

Countless marine species suffer.

Restore the otters, and entire ecosystems begin recovering.

One furry mammal floating on its back somehow influences an underwater forest.

You couldn't invent a better story.

Nature routinely outperforms fiction.

This interconnectedness should make us humble.

Instead, it often makes us grateful.

The more we learn, the less we can pretend that any species exists merely for our convenience.

Mosquitoes may test our patience.

Snakes occasionally test our courage.

Vultures rarely win popularity contests.

Yet each performs work within ecosystems that contributes to balances we scarcely understand.

I confess I still negotiate cautiously with mosquitoes.

Humility has limits.

One summer evening I was sitting on the porch when a spider began weaving a web between two railings.

Ordinarily I might have brushed it aside.

That evening I watched.

The spider moved with extraordinary precision.

Anchor.

Bridge.

Frame.

Spiral.

Strengthen.

Adjust.

The entire structure emerged from instinct refined through millions of years.

The next morning the web shimmered with dew.

Tiny droplets transformed engineering into art.

By afternoon the web had captured several insects.

The spider ate.

The mosquitoes became fewer.

I remained grateful to both biology and architecture.

There is something profoundly healthy about recognizing that we are not outside these relationships.

We participate in them every day.

The apple at breakfast required soil, sunlight, rain, bees, microbes, fungi, and farmers.

The loaf of bread represents thousands of biological and human partnerships.

Even the coffee beside me began as a flowering plant visited by pollinators before traveling halfway around the world.

Our meals are ecological stories.

We simply tend to read them too quickly.

The older I become, the more often I pause before eating.

Not out of solemn obligation.

Out of appreciation.

Someone planted.

Someone harvested.

Someone transported.

Someone prepared.

Long before that, worms loosened soil.

Fungi recycled nutrients.

Rain fell.

Sunlight arrived.

Everything I needed was already being quietly assembled by countless participants I will never meet.

Gratitude begins looking remarkably practical once you understand ecology.

One of the greatest misconceptions about humanity is that we stand apart from nature.

Air quickly corrects that misunderstanding.

Try holding your breath.

Your relationship with the atmosphere suddenly becomes very personal.

Drink a glass of water.

Thank a watershed.

Eat a peach.

Thank an orchard.

Sit beneath a tree on a hot afternoon.

Thank photosynthesis.

We have never been separate.

We have simply become distracted enough to imagine that we were.

This chapter would feel incomplete without mentioning death.

There is no graceful way around it.

Nature certainly doesn't avoid the subject.

Autumn leaves fall.

Old trees eventually topple.

Salmon die after spawning.

Wildflowers bloom for days, then fade.

The forest loses individuals while preserving the community.

As a younger man, I found this melancholy.

As an older man, I find it deeply meaningful.

Mortality gives urgency to beauty.

Cherry blossoms are precious because they do not linger.

Fireflies become magical precisely because their season is brief.

The first snowfall delights us because it cannot last forever.

Impermanence sharpens appreciation.

Perhaps that is true of our own lives as well.

The Earth has been practicing this rhythm since long before humans arrived.

Nothing lasts.

Everything matters.

Those ideas are not contradictory.

They complete one another.

A forest doesn't mourn autumn because leaves fall.

It prepares for spring.

Not because loss is unimportant.

Because renewal has always followed.

Nature refuses both despair and denial.

It acknowledges endings while continuing to create beginnings.

There is wisdom there.

Perhaps more than we realize.

Years ago, I visited a nature center with a group of children.

One little boy discovered a feather lying on the trail.

He picked it up as though he had found buried treasure.

"Whose feather is this?" he asked.

The naturalist smiled.

"It belonged to a bird."

"Did the bird die?"

"Maybe."

"Or maybe it simply lost a feather while flying."

The boy nodded thoughtfully.

Then he looked around.

"So... the bird left me a present?"

I have thought about that conversation many times.

His conclusion wasn't scientifically precise.

It was spiritually accurate.

Nature is constantly leaving gifts.

A feather.

An acorn.

A shell.

A birdsong.

A mushroom after rain.

A cloud shaped like a sailing ship.

A fox crossing the trail just before sunset.

The question is not whether the gifts are being offered.

The question is whether we arrive attentive enough to receive them.

Everything eats.

Everything is eaten.

Everything gives.

Everything receives.

Everything belongs.

That is not merely the structure of an ecosystem.

It is one of the oldest stories on Earth.

And once you begin to understand it, gratitude stops feeling like good manners.

It begins feeling like the only reasonable response to borrowing a planet where every living thing is both a beneficiary and a benefactor in the great conversation of life.

We are not standing outside that conversation.

We never were.

We have been invited to participate from the very beginning.

Chapter Ten

How Evolution Produces Miracles Every Tuesday

I once spent fifteen minutes watching a hummingbird argue with a bee.

The hummingbird believed the red trumpet vine belonged exclusively to hummingbirds.

The bee appeared unconvinced.

There was no shouting.

No legal paperwork.

No Homeowners' Association.

Just two determined little creatures negotiating access to breakfast with remarkable professionalism.

Eventually the hummingbird darted away in pursuit of another flower.

The bee continued working as though nothing unusual had happened.

The flowers, meanwhile, quietly accomplished exactly what they had intended all along.

Both visitors carried pollen to the next blossom.

Nature has an extraordinary talent for solving several problems with one elegant idea.

That little encounter lasted perhaps thirty seconds.

Behind it lay more than three billion years of life's patient experimentation.

Three billion years. It is almost impossible to think on that scale.

Human beings struggle to remember where we left our reading glasses.

Deep time asks us to imagine oceans appearing and disappearing.

Mountain ranges rising and eroding away.

Continents drifting across the globe.

Countless species emerging, changing, adapting, flourishing, and eventually giving way to others.

Not because anyone planned it.

Because life never stopped trying.

The older I become, the more I admire persistence.

Not dramatic persistence.

Quiet persistence.

The kind that keeps showing up.

The seed that tries again every spring.

The river that patiently shapes stone.

The forest that returns after fire.

The bird that builds another nest after the first one falls in a storm.

Life is astonishingly stubborn.

That stubbornness has been one of Earth's greatest strengths.

Evolution is sometimes described as survival of the fittest.

Unfortunately, many people stop listening after the word fittest and imagine the strongest, the biggest, or the most aggressive.

Nature smiles politely.

Then points to lichens growing on bare rock.

Or tiny wildflowers blooming in sidewalk cracks.

Or microscopic plankton producing much of the oxygen we breathe.

Fitness, in nature, usually means something much quieter.

It means fitting.

Fitting into changing conditions.

Fitting into relationships.

Fitting into an environment that never stops changing.

The oak survives because it belongs in its forest.

The cactus survives because it belongs in its desert.

The penguin does not complain about Antarctica.

The camel does not request more snow.

Every creature carries a long history of adaptation written into its body.

That history is worth admiring.

Take a feather.

At first glance it appears simple.

Light.

Soft.

Elegant.

Then look closer.

Each feather contains a central shaft with hundreds of delicate branches, each of which carries tiny hooklets that zip together with astonishing precision.

A bird preens because it is quietly repairing one of the finest pieces of natural engineering ever developed.

Or consider the dragonfly.

Long before humans invented helicopters, dragonflies were performing impossible aerial maneuvers with four independently controlled wings.

They can hover.

Accelerate.

Reverse direction.

Catch mosquitoes in midair with remarkable efficiency.

If engineers had designed dragonflies first, we would still be congratulating ourselves.

Nature simply built them and moved on to the next project.

Every species represents a solution.

Every adaptation tells a story.

The giraffe's neck.

The owl's silent flight.

The octopus's camouflage.

The whale's migration.

The maple seed spinning through autumn air like a tiny helicopter.

Each one whispers the same sentence.

"Life has been paying attention for a very long time."

One of the greatest mistakes we make is assuming evolution produced perfection.

It didn't.

It produced possibility.

Nature tinkers.

Experiments.

Adjusts.

Improvises.

Some ideas work beautifully.

Others disappear.

The process is less like an architect following blueprints and more like an inventor cheerfully filling a workshop with prototypes.

As someone who has spent much of his life teaching creativity, I find this wonderfully encouraging.

Failure is not merely tolerated in nature.

It is expected.

Most acorns never become oaks.

Most maple seeds never become trees.

Many nests fail.

Many hatchlings never reach adulthood.

Countless organisms begin journeys they will not finish.

Yet life continues.

Not because every attempt succeeds.

Because enough of them do.

There is wisdom there for human beings.

We often become discouraged when our first attempt doesn't work.

Nature quietly asks,

"First attempt?"

The forest has been revising itself for hundreds of millions of years.

Perhaps we can allow ourselves another try.

One spring I watched a pair of robins build a nest on top of a porch light.

From the beginning, it was an unfortunate real estate decision.

The light became warm every evening.

Rain blew in from one side.

The nest sat precariously close to the edge.

Predictably, after a violent storm, the nest fell.

The next morning I expected the robins to disappear.

Instead, they started over.

Different location.

Stronger support.

Better shelter.

No committee meeting.

No self-help books.

No lengthy discussion about confidence.

Just another attempt.

Sometimes resilience is remarkably uncomplicated.

Keep building.

Evolution has been doing exactly that.

Over immense stretches of time, life keeps trying new arrangements.

Not because the old ones were failures.

Because changing worlds require changing solutions.

That realization has made me more comfortable with change in my own life.

The world has never been finished.

Neither have I.

There is another lesson hidden in evolution that I find deeply humbling.

Every living creature on Earth shares ancestry with every other.

The maple tree.

The blue whale.

The monarch butterfly.

The mushroom beneath the log.

The chickadee at your feeder.

The dog asleep beside your chair.

The person reading these words.

Different branches.

One unimaginably ancient family tree.

Science has revealed something philosophers and poets long suspected.

Life is not a collection of strangers.

It is an extended family.

That does not erase differences.

It places them in context.

Perhaps that is why I find it increasingly difficult to think of nature as "other."

The bird outside my window is not an outsider visiting my world.

We are both temporary participants in the same astonishing story.

The same sunlight warms us.

The same gravity holds us.

The same planet carries us through space at extraordinary speed while politely refusing to mention the matter.

Humility often begins with remembering that we did not create the stage.

We merely joined the performance.

One of my favorite questions to ask while walking outdoors is wonderfully simple.

"Why this?"

Why are maple leaves shaped this way?

Why do woodpeckers have stiff tail feathers?

Why does moss prefer the shady side of the tree?

Why do geese fly in that familiar V?

Every answer uncovers another layer of history.

Another adaptation.

Another elegant solution.

The world becomes richer every time curiosity asks another question.

I have come to believe that evolution's greatest gift is not simply the diversity of life.

It is the diversity of wonder.

There is always another marvel waiting just beyond what we already know.

The scientist discovers a new species.

The child discovers a ladybug.

The gardener notices the first tomato flower.

The birder hears a warbler returning in spring.

The astronomer finds another planet.

The naturalist kneels beside an unfamiliar mushroom.

Different scales.

The same delight.

Wonder does not require novelty.

It requires attention.

The Earth continues producing miracles every single day.

Not because the laws of nature are suspended.

Because those laws are far more imaginative than we once believed.

A seed becoming a forest.

A caterpillar becoming a butterfly.

A salmon finding the stream where it was born.

A spider weaving geometry from silk.

A bee navigating by the position of the sun.

A child asking, "Why?"

These are not interruptions in the ordinary course of life.

They are the ordinary course of life.

We simply become accustomed to them.

That may be the greatest danger of familiarity.

It persuades us that because something happens every day, it is somehow less miraculous.

The sunrise suffers from this problem.

So do birds.

So do trees.

So do people.

The graybeard eventually learns a delightful secret.

Common is not the opposite of miraculous.

Common is often where miracles choose to hide.

The robin on your lawn is the product of an unbroken chain of life stretching back through unimaginable ages.

The oak shading your street represents millions of years of quiet experimentation.

The hummingbird arguing with the bee carries within it a history older than mountains.

And you–

The curious soul holding this book–

are one more participant in that extraordinary unfolding.

Not standing above it.

Not standing apart from it.

Standing gratefully within it.

Borrowing, for a little while, a planet that has spent billions of years becoming beautiful enough to astonish anyone willing to stop and look.

Chapter Eleven

Nobody Falls in Love with Statistics

I have spent a fair portion of my life standing in front of classrooms.

That experience teaches you many useful things.

For example, never begin a lecture immediately after lunch if your first slide contains seventeen bullet points.

Human digestion is a formidable educational opponent.

I've also learned that statistics are peculiar creatures.

They are extraordinarily important.

They help us understand disease, poverty, economics, education, wildlife populations, climate, and almost everything else that benefits from careful measurement.

Statistics save lives.

They guide policy.

They expose problems that anecdotes alone cannot reveal.

They deserve respect.

They are also remarkably poor at making people cry.

No one has ever interrupted a dinner conversation to say,

"Last night I read a bar graph that changed my life."

People remember stories.

We carry them for decades.

We tell them to our grandchildren.

We quote them at weddings.

We repeat them around campfires.

Numbers inform the mind.

Stories recruit the heart.

The best environmental leaders have always understood this.

Rachel Carson did not simply write about pesticide concentrations.

She invited readers to imagine a spring without birdsong.

Aldo Leopold did not argue only for better land management.

He wrote about watching a dying wolf and realizing that mountains might understand something human beings had overlooked.

Robin Wall Kimmerer teaches botany through gratitude.

David Attenborough introduces viewers to individual animals before asking us to care about entire ecosystems.

Again and again, the pattern repeats.

First, relationship.

Then responsibility.

Science tells us what is happening.

Story tells us why it matters.

Years ago, I visited a small nature center where an elderly volunteer was leading children along a woodland trail.

She stopped beside an enormous white pine.

"How old do you think this tree is?" she asked.

The guesses ranged from twenty years to approximately the age of dinosaurs.

Children possess admirable confidence.

She smiled.

"About one hundred and fifty years."

The group grew quiet.

Then she said something I have never forgotten.

"This tree was already here when your great-great-grandparents were children."

Immediately the tree changed.

Not physically.

Emotionally.

It was no longer simply a tall plant.

It had become a witness.

The children reached out to touch the bark.

One little boy wrapped both arms around the trunk.

A girl looked upward for what seemed like several minutes.

Nobody instructed them to care.

Perspective did the work.

The tree had acquired a story.

That afternoon reminded me that people rarely become attached to abstractions.

We become attached to places.

The creek where we caught our first fish.

The trail where our father taught us to identify birds.

The apple tree in our grandmother's yard.

The pond where we watched tadpoles become frogs.

The old maple that shaded neighborhood conversations every summer.

Love usually has an address.

Psychologists have a phrase for this.

Place attachment.

It describes the emotional bonds people develop with landscapes, neighborhoods, parks, lakes, forests, and even particular trees.

These attachments influence behavior more than we often realize.

People volunteer to restore streams they grew up beside.

Communities fight to preserve beloved parks.

Neighbors plant flowers around vacant lots because they remember what those places once meant.

The motivation rarely begins with policy.

It begins with affection.

The older I become, the less interested I am in asking,

"How do we convince people to care?"

I find myself asking a different question.

"What helped them fall in love in the first place?"

Because once affection exists, many other behaviors follow naturally.

You don't have to remind gardeners to water roses.

You don't have to persuade birdwatchers to preserve habitat.

You don't have to convince grandparents to protect the playground where their grandchildren laugh.

Love is astonishingly self-motivating.

The challenge is that modern life often introduces us to nature primarily through problems.

Climate change.

Habitat loss.

Pollution.

Species decline.

These realities matter deeply.

Ignoring them would be irresponsible.

But if someone's first encounter with the natural world is a list of crises, they may conclude that nature is primarily something broken.

That would be a tragedy.

Because before the Earth became vulnerable, it was beautiful.

Before ecosystems became threatened, they became beloved.

The order matters.

Children seldom fall in love with forests because someone assigned a chapter on deforestation.

They fall in love because they built forts beneath trees.

They caught fireflies.

They found salamanders beneath logs.

They listened to owls after dark.

Experience writes emotional memories that information later strengthens.

I remember a professor who once held up a leaf during class.

Just one leaf.

Instead of explaining photosynthesis immediately, he asked us to look.

Really look.

Veins branching through green tissue.

Tiny imperfections.

Edges shaped by genetics and weather.

Light passing through the surface.

For nearly five minutes he said almost nothing.

Then he quietly remarked,

"Every one of these is a solar-powered food factory."

The sentence landed differently because observation came first.

We had already begun admiring the leaf before learning how astonishing it actually was.

Good teachers understand timing.

Nature does, too.

One of my favorite examples comes from the National Parks.

Ask visitors why they return.

Very few begin with geological statistics.

Instead they describe standing at the rim of a canyon while the sunrise transformed stone into fire.

Watching elk cross a meadow.

Hearing wolves howl.

Smelling pine after rain.

Feeling very small beneath ancient sequoias.

Emotion arrives before explanation.

Wonder precedes understanding.

Neither replaces the other.

They strengthen one another.

This is true beyond nature.

Consider the people you love most.

You probably cannot explain precisely why you love them.

Certainly not with spreadsheets.

Imagine proposing marriage by presenting a carefully prepared pie chart.

"Ladies and gentlemen, as you can clearly see, compatibility has increased by 14.6 percent over the previous fiscal year."

The wedding would be memorable for all the wrong reasons.

Love resists quantification.

Nature does, too.

That doesn't diminish science.

It reminds us that human beings possess both minds and hearts.

The mind asks,

"Is this true?"

The heart asks,

"Does this matter?"

Wisdom insists on answering both.

I sometimes worry that our culture mistakes information for transformation.

If knowledge alone changed behavior, every physician would exercise perfectly.

Every financial advisor would retire wealthy.

Every marriage counselor would enjoy flawless relationships.

Life is more complicated.

People change when understanding becomes personal.

The environmental movement is no exception.

A graph may reveal declining pollinator populations.

A child watching a monarch butterfly emerge from its chrysalis creates something entirely different.

Now the graph has a face.

Now the issue has wings.

Now conservation has become personal.

The remarkable thing is that nature keeps providing opportunities for these encounters.

Every day.

Every season.

A turtle crossing a road.

A fox slipping through morning fog.

A cardinal bright against fresh snow.

A seed becoming a sunflower.

A spider web sparkling with dew.

Tiny invitations to affection.

The Earth has never stopped introducing itself.

We have simply become busy during the introductions.

The older I grow, the more convinced I become that environmental stewardship is less about changing minds than about reopening eyes.

Because people protect what they know.

They know what they notice.

They notice what they have learned to love.

And nobody–not ever–falls in love with statistics alone.

We fall in love with places.

With stories.

With living things whose names we know.

With mornings spent beside quiet lakes.

With old trees that remember generations.

With birds that return every spring as faithfully as old friends.

We fall in love with a planet that, despite all its troubles, continues offering beauty with extravagant generosity.

Statistics tell us what we stand to lose.

Wonder reminds us why it is worth saving.

The Earth needs both.

But if we must choose where to begin, I believe the better starting place is still the oldest one.

Go outside.

Meet your neighbors.

Learn a few names.

Listen carefully.

The numbers will eventually make sense.

Love usually arrives first.

Chapter Twelve

The Porch Light Theory of Conservation

Every evening, just before dusk, an elderly couple a few streets over turns on their porch light.

Not because they expect company.

Not because they are worried about burglars.

Not because the bulb is connected to some sophisticated smart-home system that has been thoughtfully synchronized with the position of the sun and the phases of the moon.

No.

They simply believe that a light should be on when darkness arrives.

That's all.

The house itself isn't particularly remarkable.

White siding.

Blue shutters.

A swing on the front porch.

Flower boxes that somehow look cheerful even after a week without rain.

Yet every time I walk past in the evening, I notice the light.

It changes the entire house.

The porch becomes less like architecture and more like an invitation.

Someone is home.

Someone has prepared a place.

Someone has quietly announced to the world,

"You're welcome here."

I have come to believe that conservation begins with exactly that impulse.

Not ownership.

Hospitality.

For many years, I thought environmental stewardship meant protecting places that were far away.

National parks.

Remote wilderness.

Rainforests.

Mountain ranges whose names I occasionally mispronounce.

Those places matter enormously.

They deserve every effort we can make to preserve them.

But somewhere along the way, I realized something embarrassingly obvious.

Most living things don't reside in national parks.

They live in neighborhoods.

Backyards.

Roadside ditches.

Schoolyards.

Community gardens.

Vacant lots.

Farm fields.

Streams running behind shopping centers.

Patches of milkweed beside railroad tracks.

Nature has never limited itself to postcard destinations.

It has always accepted ordinary addresses.

That realization changed the kinds of questions I began asking.

Not,

"How do we save the wilderness?"

But,

"How do we make more places feel welcoming to life?"

The difference is subtle.

It is also profound.

Imagine inviting friends to dinner.

You don't merely unlock the front door.

You prepare.

You set the table.

You sweep the porch.

You make room.

Hospitality is active.

The Earth seems to appreciate the same courtesy.

A homeowner plants native flowers.

Soon butterflies arrive.

Someone adds a birdbath.

Songbirds become regular visitors.

A community restores a neglected creek.

Fish gradually return.

A school plants trees.

Children inherit shade they did not have the year before.

No grand speeches.

No dramatic announcements.

Simply a series of small invitations.

"You're welcome here."

One of the most hopeful discoveries in modern ecology is that nature responds remarkably well when given the opportunity.

Not always immediately.

Not perfectly.

But consistently enough to inspire optimism.

Restore wetlands, and birds often return.

Remove invasive plants, and native wildflowers begin reclaiming space.

Plant milkweed, and monarch butterflies notice.

Install nesting boxes, and bluebirds sometimes move in before you've finished admiring your own handiwork.

Life seems to have been waiting for someone to leave the porch light on.

This is one of my favorite characteristics of the natural world.

It wants to recover.

Forests regrow.

Prairies bloom again.

Streams clear.

Wildlife returns.

Given half a chance, creation displays astonishing resilience.

That doesn't mean damage isn't real.

It is.

Some losses cannot be undone.

Some species, once gone, leave silences no amount of wishing can fill.

Rachel Carson understood this.

So did Aldo Leopold.

So do today's conservation biologists.

Hope should never become an excuse for carelessness.

But neither should realism become an excuse for despair.

The Earth has survived ice ages, volcanic eruptions, asteroid impacts, shifting continents, and countless other upheavals.

Its resilience is extraordinary.

Our responsibility is to avoid making its work unnecessarily harder.

Hospitality extends beyond wildlife.

Healthy neighborhoods welcome people, too.

Think about your favorite park.

Perhaps it has shaded benches.

Children laughing near a playground.

Birds singing overhead.

A walking trail winding beneath old trees.

Someone reading a book.

Someone teaching a grandchild to skip stones.

Someone quietly eating lunch beneath an oak.

Notice what happens in places like these.

Nature and community stop feeling like separate subjects.

The same tree providing shade for a family picnic is sheltering a nesting robin.

The same rain garden filtering stormwater is filled with bees.

The same trail encouraging exercise also protects habitat.

Good places rarely choose between human flourishing and ecological flourishing.

They accomplish both.

Nature has never insisted that we stay away.

It has always invited us to belong.

That realization has softened my understanding of conservation.

When I was younger, I imagined stewardship mostly as restraint.

Don't cut the tree.

Don't pollute the river.

Don't disturb the habitat.

Those remain wise principles.

But stewardship also includes positive acts.

Plant.

Restore.

Welcome.

Share.

Teach.

Protect.

Celebrate.

Enjoy.

The Earth deserves more than our guilt.

It deserves our participation.

Years ago, I met a retired gentleman whose backyard had become famous among neighborhood children.

Not because he owned expensive equipment.

Quite the opposite.

He had simply planted native flowers, allowed a corner of the yard to remain a little wild, installed several bird feeders, and built a small pond no larger than a dining room table.

By suburban standards, it was modest.

By ecological standards, it was extraordinary.

Dragonflies patrolled the pond.

Tree frogs appeared each spring.

Goldfinches visited the coneflowers.

Monarch butterflies drifted through in late summer.

Children arrived carrying field guides instead of video game controllers.

He laughed when people complimented him.

"I didn't create any of this," he said.

"I just stopped making it difficult."

I've thought about those words many times.

Perhaps conservation is less about controlling nature than cooperating with it.

The world already knows how to grow forests.

It knows how to pollinate flowers.

It knows how to build soil, clean water, and welcome spring.

Our task is often remarkably simple.

Stop getting in the way.

Of course, that simplicity should not fool us.

Small actions accumulate.

A single flower may feed a bee.

A hundred gardens become habitat.

One restored stream becomes a healthier watershed.

One neighborhood planting trees influences temperatures, birds, insects, rainfall, and quality of life for generations.

The porch light shines farther than we imagine.

That may be true of kindness as well.

Every act of hospitality spreads.

One neighbor notices your garden and plants one of their own.

Children raised among butterflies grow into adults who value butterflies.

Communities rediscover neglected parks.

People who experience beauty often become its protectors.

The light moves from porch to porch.

From yard to yard.

From heart to heart.

There is another reason I love porch lights.

They make it easier to find your way home.

Late one evening, years ago, after driving much longer than planned, I turned onto my street tired and ready for sleep.

Long before I recognized my own house, I saw the warm glow of the porch light waiting at the end of the driveway.

Nothing dramatic.

Just enough light to say,

"You're home."

I wonder whether the Earth has been trying to tell us the same thing all along.

Not,

"This belongs to you."

But,

"You belong here."

There is an enormous difference.

Ownership often asks,

"What can I take?"

Belonging asks,

"How can I care for this place?"

Ownership counts possessions.

Belonging builds relationships.

Ownership ends with a deed.

Belonging begins with gratitude.

The older I become, the less interested I am in owning pieces of the world.

I would rather leave porch lights behind.

Gardens that feed butterflies.

Trees whose shade I may never enjoy.

Trails that invite strangers to walk more slowly.

Creeks that run a little cleaner.

Children who know the names of birds.

Neighbors who pause to admire the evening sky.

Small lights.

Quiet invitations.

Gentle reminders that this remarkable planet has always been more than a resource.

It is our shared neighborhood.

And every borrowed neighborhood becomes a little brighter when someone remembers to leave the porch light on.

Chapter Thirteen

Children Already Know This

Several years ago, I was walking through a park with my granddaughter when she stopped so abruptly that I nearly walked into her.

She didn't say anything.

She simply pointed.

There, inching its way across the sidewalk with magnificent determination, was a woolly bear caterpillar.

Now, I had seen woolly bears before.

Thousands, probably.

As a child.

As an adult.

Crossing sidewalks.

Crossing roads.

Climbing weeds.

Doing whatever mysterious business woolly bears conduct during their surprisingly busy lives.

Most of the time, I acknowledged them with the brief mental note we reserve for familiar things.

"Oh look... a caterpillar."

Then I continued walking.

My granddaughter had a different agenda.

She crouched.

Then she lay flat on the sidewalk.

Not because she had fallen.

Because she wanted to see better.

For the next ten minutes, we watched that little caterpillar climb over pebbles that, from its perspective, were respectable mountains.

It explored cracks in the pavement as though evaluating possible real estate.

It occasionally stopped, apparently to reconsider its route or perhaps simply because caterpillars have earned the right to move at whatever speed they find appropriate.

Eventually, my granddaughter looked at me and whispered,

"He's trying really hard."

She had quietly done something adults often forget to do.

She had granted significance to a very small life.

Children possess this remarkable ability.

A ladybug is never just a ladybug.

A frog is not just a frog.

A puddle is not merely accumulated rainwater.

A stick is not dead wood.

Everything is alive with possibility.

Everything deserves investigation.

Everything has a story waiting to be discovered.

Adults often describe this as innocence.

I'm not so sure.

I think it's attention.

Children haven't yet learned which things society considers too ordinary to notice.

And so they notice nearly everything.

Watch a group of young children on a nature walk.

The adults usually begin by looking outward.

The children immediately begin looking downward.

They discover mushrooms the size of marbles.

Acorns with tiny holes.

Feathers.

Snail shells.

Oddly shaped leaves.

Insects that no adult in the group had seen.

Children live at a different altitude.

Perhaps wisdom requires spending more time there.

One of my favorite memories comes from leading a group of elementary school students through a botanical garden.

We had barely entered the gate when one little boy shouted,

"Look!"

Every adult immediately lifted their eyes toward the trees.

Nothing.

The child was pointing at the ground.

A line of ants.

Dozens of them.

Each carrying fragments of leaves several times larger than themselves.

The children gathered in a circle.

The adults followed.

Within minutes, the planned lesson had disappeared.

Instead of discussing the rare plants we had intended to study, we spent half an hour watching ants solve transportation problems with astonishing efficiency.

Nobody complained.

Least of all the children.

Nature had rewritten the lesson plan.

I've often thought education would improve if we allowed curiosity to interrupt us more frequently.

The older I become, the less interested I am in asking,

"What should we teach children about nature?"

I find myself asking,

"What should children remind us about nature?"

The answers arrive surprisingly quickly.

Slow down.

Look closer.

Ask why.

Touch carefully.

Laugh often.

Don't be embarrassed by wonder.

Celebrate discoveries.

Return tomorrow to see what changed.

These are not childish habits.

They are scientific habits.

The finest naturalists in history shared them.

The greatest ecologists never stopped asking questions that sounded remarkably like those of a curious seven-year-old.

"Why does this happen?"

"What lives under that log?"

"What happens if...?"

Children and scientists are separated by vocabulary.

Not by curiosity.

Somewhere along the road to adulthood, many of us begin confusing knowledge with familiarity.

We think that because we know what something is called, we have finished learning about it.

"That's a robin."

Conversation over.

Children disagree.

"What does it eat?"

"Where does it sleep?"

"Does it have babies?"

"Why is its chest orange?"

"Can it remember my face?"

One answer simply creates three new questions.

That is not ignorance.

That is intellectual generosity.

Curiosity keeps the door open.

Certainty quietly closes it.

Years ago, I visited a preschool where the teachers had transformed an ordinary corner of the playground into what they called the Discovery Garden.

There were no expensive exhibits.

No digital displays.

No elaborate sculptures.

Just native flowers.

Logs.

Stones.

A shallow pond.

Bird feeders.

A patch of tall grass.

The children treated the place as though they had been given access to another planet.

Every day brought announcements.

"A butterfly!"

"A frog!"

"The flowers opened!"

"The bird is back!"

"The mushrooms are bigger today!"

Adults, meanwhile, might have walked through without noticing anything unusual.

The garden had not changed.

The observers had.

One little girl proudly introduced me to a dandelion.

"This one's my friend."

I smiled.

"Does it have a name?"

She thought carefully.

"I don't think he told me yet."

Children have a marvelous habit of assigning personality to the living world.

Adults call this anthropomorphism.

Scientists gently caution against assuming that trees have feelings exactly like ours or that squirrels spend afternoons contemplating philosophy.

The caution is wise.

But I also wonder if, in correcting children's biology, we sometimes accidentally weaken their empathy.

There is value in treating living things as worthy of respect long before we understand every detail about them.

After all, good manners often arrive before complete knowledge.

You don't need a degree in ornithology to avoid disturbing a nesting bird.

You don't need a doctorate in botany to appreciate an old oak.

Respect does not require expertise.

It requires attention.

The environmental challenges facing the next generation are real.

Climate.

Habitat loss.

Pollution.

Biodiversity.

These issues deserve thoughtful education.

But I hope we never make the mistake of introducing children to nature primarily through catastrophe.

The Earth should first be introduced as a friend.

Not as a patient in intensive care.

Let children catch lightning bugs before teaching them abcut declining insect populations.

Let them build forts beneath trees before discussing deforestation.

Let them fall in love with streams before asking them to restore rivers.

Affection creates resilience.

Despair rarely does.

The remarkable thing is that children seem predisposed toward affection.

They don't require convincing.

Only opportunities.

Take them outside.

Then mostly stay out of the way.

One spring afternoon I watched a little boy spend nearly twenty minutes trying to help an earthworm cross a sidewalk before the sun dried it out.

His engineering strategy was questionable.

His compassion was flawless.

Nobody had assigned the task.

Nobody was grading the exercise.

He simply believed the small creature mattered.

There, in the middle of an ordinary afternoon, was conservation in its purest form.

Not legislation.

Not policy.

Not fundraising.

Compassion meeting opportunity.

I wonder how different our world might become if adults practiced that instinct a little more often.

Children eventually grow older.

They learn schedules.

Responsibilities.

Budgets.

Deadlines.

Taxes.

Passwords.

Somewhere along the journey, many stop lying on sidewalks to watch caterpillars.

That is understandable.

Life becomes busy.

But perhaps becoming wise is not merely adding new knowledge.

Perhaps wisdom also involves recovering forgotten habits.

Kneeling beside wildflowers.

Looking under rocks.

Learning the names of birds.

Watching clouds.

Standing quietly beneath old trees.

Feeling genuine excitement over the return of spring.

The child inside us never entirely disappears.

He merely gets buried beneath calendars.

She becomes hidden beneath obligations.

Nature has an extraordinary way of introducing us to them again.

One feather.

One mushroom.

One curious question.

One woolly bear crossing a sidewalk with unreasonable confidence.

I still think about that caterpillar.

Not because it accomplished anything remarkable.

It probably reached the grass, became a moth, and quietly continued the business of being a woolly bear.

I think about it because of what happened to me.

For ten minutes, the world slowed down.

A sidewalk became an expedition.

A caterpillar became a teacher.

A granddaughter became my guide.

Children already know what many adults spend years trying to rediscover.

The world is alive.

Everything is interesting.

Nothing is truly ordinary.

And every creature, no matter how small, deserves at least a moment of our undivided attention.

That may be one of the finest instructions anyone could receive while borrowing a planet.

Chapter Fourteen

Your Yard Is Bigger Than You Think

I have never owned a rainforest.

I've checked.

The deed is disappointingly specific.

House.

Garage.

A modest patch of lawn that insists on growing faster whenever I have other plans.

Several flower beds that occasionally resemble gardens and occasionally resemble botanical negotiations.

One maple tree.

One stubborn patch of moss that has consistently ignored my landscaping ambitions.

That's about it.

If you were to look at my property from an airplane, you would probably miss it.

If you were looking for ecological significance, you would certainly miss it.

For years, so did I.

Like many people, I believed conservation happened somewhere else.

In Yellowstone.

In the Amazon.

In vast wildlife refuges with impressive visitor centers and gift shops selling educational refrigerator magnets.

Those places matter enormously.

But one afternoon, while talking with a conservation biologist, I heard a sentence that quietly rearranged my understanding of the world.

He smiled and said,

"Most wildlife doesn't live in national parks."

I nodded.

Then I paused.

He was right.

Most birds spend far more time in neighborhoods than in wilderness.

Most pollinators don't distinguish between a famous botanical garden and your flower bed.

A monarch butterfly has never refused to land on milkweed because the address wasn't prestigious.

A chickadee doesn't care whether your maple appears in travel brochures.

Nature has never been particularly impressed by zip codes.

That realization delighted me.

It also removed one of my favorite excuses.

I could no longer admire conservation from a comfortable distance.

It had quietly moved next door.

One spring, I decided to replace a small patch of decorative plants with native wildflowers.

Nothing dramatic.

The gardening equivalent of rearranging the furniture.

The first year looked... underwhelming.

A few green shoots.

Some hopeful leaves.

Several weeds that displayed considerably more enthusiasm than the flowers.

I briefly wondered whether I had misunderstood the instructions.

Then came the second spring.

The flowers bloomed.

Soon after, the bees arrived.

Not one or two.

Dozens.

Bumblebees.

Tiny native bees.

Butterflies.

Hoverflies.

Creatures I had spent decades overlooking suddenly treated my yard like a neighborhood café.

Apparently, word had spread.

One afternoon I stood quietly with a cup of coffee, watching the flowers sway with life.

It occurred to me that I hadn't merely planted flowers.

I had opened a restaurant.

Nature had accepted the invitation.

That is one of the great pleasures of gardening.

You discover very quickly that the garden isn't really for you.

You're simply one of the guests.

The bees arrive.

The birds investigate.

Spiders establish neighborhoods.

Toads move into the cool shade beneath broad leaves.

Dragonflies patrol overhead.

Even rabbits, despite occasionally expressing opinions about my lettuce, remind me that I am participating in something larger than landscaping.

Ecologists have a wonderful phrase for this.

Habitat.

When most people hear the word, they imagine remote wilderness.

Old-growth forests.

Wetlands.

Prairies stretching to the horizon.

Those are habitats.

So is your backyard.

So is the vacant lot down the street.

So is the flowering tree beside the library parking lot.

Habitat simply means a place where life can live.

That is an encouraging definition.

It suggests that almost anyone can create more of it.

Years ago, entomologist Douglas Tallamy began sharing a simple but revolutionary idea.

If millions of homeowners each made small changes to support native plants and wildlife, the cumulative effect could rival many large conservation projects.

At first glance, this seems improbable.

Until you remember that landscapes are mosaics.

One yard touches another.

One neighborhood connects to another.

One restored stream eventually joins a river.

Small pieces create large patterns.

Nature has always understood this.

Standing in my yard one summer morning, I watched a goldfinch land on a purple coneflower.

It stayed only a few moments before flying across the street into my neighbor's garden.

Later, a monarch drifted through both yards before continuing toward the nearby park.

Neither bird nor butterfly recognized our property line.

The surveyor had.

The wildlife had not.

Perhaps that should tell us something.

The Earth acknowledges relationships more readily than boundaries.

Another lesson arrived through an unexpected teacher.

Leaves.

For years I treated autumn leaves as opponents.

Every October they descended upon the lawn with astonishing determination.

I raked.

They returned.

I raked again.

The maple appeared unconcerned.

Then a naturalist explained something wonderfully obvious.

Those leaves weren't litter.

They were next year's soil.

The forest had been making compost for hundreds of millions of years before humans invented leaf bags.

Now, I still rake some leaves.

Neighborhood expectations have their place.

But I also leave many beneath shrubs and trees.

Fireflies spend part of their lives there.

Butterflies overwinter there.

Countless insects find shelter there.

The leaves quietly become nourishment.

The yard becomes a little more alive.

And I become a little less convinced that tidiness is always the same thing as health.

Nature occasionally prefers messy.

Not neglected.

Alive.

There is an important distinction.

One of my neighbors grows tomatoes every summer.

Another keeps bees.

Someone else has transformed a corner of their yard into a tiny prairie.

A retired couple down the block planted an oak after their first grandchild was born.

Without planning it, our street has become something of an ecological neighborhood.

Birds move between yards.

Pollinators travel from flower to flower.

Trees gradually enlarge the canopy.

Children catch butterflies they never would have seen twenty years earlier.

No committee organized it.

No strategic plan was drafted.

People simply began making small decisions.

Good ones.

Those decisions accumulated.

That's one of my favorite characteristics of hope.

It compounds.

The environmental challenges facing our world are undeniably large.

Sometimes overwhelmingly so.

When faced with enormous problems, human beings often make one of two mistakes.

We either believe our individual actions are too small to matter.

Or we assume only governments, corporations, or scientists possess meaningful influence.

Fortunately, nature is not nearly so dismissive.

A bee visits one flower at a time.

An earthworm improves one tunnel at a time.

A tree grows one ring at a time.

A mushroom decomposes one fallen branch at a time.

The forest never becomes discouraged because today's acorn isn't yet an oak.

Life trusts accumulation.

Perhaps we should, too.

I have a friend whose yard contains a weathered wooden bench beneath an old maple.

He told me he built it for birds.

I pointed out that birds rarely sit on benches.

He laughed.

"No," he said.

"I built it so I'd sit still long enough to notice the birds."

That may be the wisest landscaping advice I've ever received.

Sometimes the most important feature in a garden is not another flower.

It's a place to pay attention.

A place where coffee cools while chickadees argue.

Where butterflies drift through afternoon sunlight.

Where children search for caterpillars.

Where neighbors stop to talk.

Where gratitude quietly catches up with us.

By now, you may have noticed that this chapter has contained remarkably little advice about fertilizers, pruning schedules, or lawn equipment.

That is intentional.

This isn't really a chapter about gardening.

It's a chapter about belonging.

The yard matters because it teaches us that stewardship has an address.

It is not merely an idea discussed in conference rooms.

It is practiced in places.

One flower at a time.

One tree at a time.

One birdbath.

One patch of milkweed.

One conversation over the backyard fence.

One child discovering a praying mantis.

One neighbor deciding to plant another oak.

The older I become, the less interested I am in owning an impressive yard.

I would rather borrow a small corner of the Earth so well that life quietly flourishes there after I've gone.

Perhaps that is the true measure of stewardship.

Not the size of the property.

The generosity of the welcome.

Because your yard has never really been measured in square feet.

It has always been measured in relationships.

The bees already know that.

The butterflies do, too.

The birds have been trying to tell us for years.

We simply had to become quiet enough to hear them.

Chapter Fifteen

A Bird Feeder Is Not About Birds

The first bird feeder I ever bought was too small.

I know this because the squirrels informed me.

Immediately.

With remarkable enthusiasm.

Within forty-eight hours they had developed what can only be described as a coordinated acquisition strategy.

One climbed the shepherd's hook.

Another launched from the maple tree.

A third served, I assume, as quality control.

The birds, meanwhile, waited politely nearby, apparently accustomed to squirrels treating life as an Olympic event.

I responded the way many suburban homeowners do.

I purchased a squirrel-proof feeder.

The squirrels accepted this as a personal challenge.

I added a baffle.

They considered it an engineering problem.

I moved the feeder farther from the tree.

They recalculated.

For several weeks I was locked in an escalating technological contest with rodents possessing brains approximately the size of walnuts.

The squirrels won.

I don't mean occasionally.

I mean decisively.

Eventually I surrendered.

Not completely.

I simply accepted that a modest percentage of my birdseed budget had become a community outreach program for unusually athletic mammals.

Oddly enough, my enjoyment increased.

Once I stopped trying to control every visitor, the feeder became much more interesting.

Goldfinches arrived in bright flashes of yellow.

Cardinals appeared like living Christmas ornaments.

Chickadees darted in with cheerful confidence.

Nuthatches walked headfirst down the maple trunk, apparently convinced that gravity was more of a suggestion than a rule.

Downy woodpeckers announced themselves with purposeful determination.

House finches argued amiably over sunflower seeds.

Even the squirrels, despite their questionable ethics, contributed a certain comic value.

Morning coffee gradually moved from the kitchen table to the window overlooking the feeder.

Without intending to, I had developed a ritual.

Every day brought familiar faces.

And every day brought surprises.

One winter morning, after a heavy snowfall, a bird I had never seen before landed quietly on the feeder.

I hurried to my field guide.

Then to my binoculars.

Then back to the field guide because I had already forgotten what I thought I had identified.

Eventually I confirmed that it was a white-throated sparrow.

I felt absurdly proud.

Not because I had accomplished anything difficult.

Because the bird was no longer anonymous.

It had become an acquaintance.

Names change relationships.

The first cardinal is "a bird."

The tenth becomes the cardinal who always arrives just after sunrise.

The chickadee becomes the bold little fellow who scolds squirrels from the lilac bush.

The mourning dove becomes the patient visitor who always waits until the commotion settles.

Whether those stories exist primarily in my imagination is beside the point.

Attention has transformed strangers into neighbors.

That, I suspect, is the hidden purpose of bird feeders.

They are not primarily feeding birds.

They are feeding observers.

Before the feeder, birds occupied the edges of my awareness.

Afterward, they became part of the rhythm of my days.

I noticed the first goldfinch of spring.

The arrival of migrating warblers.

The silence after a hawk passed overhead.

The disappearance of hummingbirds in autumn.

The return of juncos announcing winter.

Without realizing it, I had begun measuring the seasons in feathers rather than calendars.

The feeder had become a clock.

One of the quiet tragedies of modern life is that many of us have lost these seasonal markers.

Our climate-controlled homes maintain nearly identical temperatures year-round.

Fresh strawberries appear in grocery stores regardless of month.

Electric lights blur the distinction between summer evenings and winter afternoons.

Nature continues keeping time.

We simply stopped consulting its calendar.

Birds remember.

The first robin of spring.

The restless flocks gathering before migration.

The growing chorus at dawn in May.

The surprising quiet of late August.

Each season has its own soundtrack.

The feeder taught me to listen.

It also taught me patience.

Birds are wonderfully indifferent to human schedules.

You cannot command a hummingbird to arrive at precisely 9:15.

Bluebirds refuse appointments.

Owls decline calendar invitations.

Nature has never synchronized itself with office hours.

Watching birds requires something increasingly rare.

Availability.

You must become still enough that life continues around you.

That stillness changes more than observation.

It changes the observer.

I have noticed that whenever visitors come to the house, conversation eventually pauses.

Someone points toward the feeder.

"What's that one?"

Immediately everyone becomes a beginner.

Even people who have never shown the slightest interest in birds lean closer to the window.

A child identifies the cardinal.

Someone mistakes a female finch for a sparrow.

Another person notices the woodpecker.

For a few minutes, the room becomes wonderfully united by curiosity.

Birds accomplish something that politics rarely manages.

They give strangers something delightful to talk about.

There is another reason I treasure the feeder.

It reminds me that generosity attracts life.

The feeder offers no guarantees.

Some mornings it remains quiet.

Other days it becomes unexpectedly busy.

Yet every sunrise begins with the same quiet invitation.

"If you need something, you're welcome here."

The porch light from the previous chapter has acquired feathers.

Ecologists often speak about habitat.

Bird lovers understand that habitat is not merely a place.

It is a relationship.

Food.

Water.

Shelter.

Safety.

A place to raise young.

A place to rest during migration.

The remarkable thing is that small acts can help create those relationships.

One feeder.

One birdbath.

One native shrub.

One old tree left standing.

Tiny invitations.

Life notices.

One afternoon I watched a father helping his young daughter refill the feeder.

She carefully poured sunflower seeds while asking questions faster than he could answer them.

"Do birds have best friends?"

"Where do they sleep?"

"How do they know where to fly?"

"Why doesn't the woodpecker get a headache?"

Excellent questions.

Questions worthy of both science and wonder.

The father admitted when he didn't know.

Then they looked up the answers together.

That may have been the greatest gift of all.

The feeder wasn't simply attracting birds.

It was producing conversations.

Curiosity was being inherited.

I have become convinced that conservation often passes from one generation to the next in moments exactly like that.

Not through lectures.

Through shared attention.

Through names.

Through stories.

Through ordinary mornings spent beside windows.

Some years ago, I met an elderly gentleman who had been feeding birds for more than forty years.

I asked him which species was his favorite.

He thought for a long moment.

Then he smiled.

"The next one."

I have never forgotten that answer.

Hope always has room for one more arrival.

One more surprise.

One more reason to look out the window.

One more invitation to pay attention.

Perhaps that is why bird feeders matter so much.

Not because the birds cannot survive without us.

Most can.

Not because they need our admiration.

They don't.

Bird feeders matter because they quietly reshape human hearts.

They teach us names.

They teach us patience.

They teach us the rhythm of seasons.

They teach us that life continues whether we notice it or not–but becomes infinitely richer when we do.

The birds, I suspect, would manage perfectly well without my feeder.

I am much less certain that I would manage as well without them.

Every morning they remind me that the world woke up before I did.

That beauty still arrives without invitation.

That songs are offered freely.

That generosity often comes with wings.

And that borrowing a planet is much easier when you begin each day by greeting your neighbors.

Even the noisy ones.

Especially the ones who occasionally steal the birdseed.

Chapter Sixteen

You Do Not Own Your Shade Tree

Several years ago, a storm brought down a large maple at the edge of our neighborhood.

It had stood there for longer than anyone on the street could remember.

Children had climbed it.

Teenagers had leaned bicycles against it.

Parents had waited beneath its shade for school buses.

Dogs had conducted the important business of neighborhood diplomacy around its trunk.

It had quietly become part of everyone's map of home.

The morning after the storm, people gathered around the fallen tree.

Nobody had organized the meeting.

No invitations had been sent.

People simply came.

Someone took photographs.

Someone counted the rings.

Someone recalled carving initials into one of its lower branches fifty years earlier.

One woman remembered sitting beneath it with her father after her mother died.

Another remembered collecting maple helicopters every autumn.

The tree had not merely occupied space.

It had occupied memory.

That is an entirely different thing.

As I listened to those conversations, it occurred to me that not one person referred to the tree as timber.

Not one.

No one estimated its market value.

No one discussed the price of firewood.

They talked about birthdays.

Neighborhoods.

Summers.

Shade.

Stories.

The tree had become something economists rarely measure.

It had become part of a shared life.

We often speak about owning land.

I understand why.

Our legal system depends upon ownership.

Property lines matter.

Titles matter.

Deeds matter.

If you've ever accidentally mowed six inches into your neighbor's lawn, you'll appreciate that boundaries occasionally deserve attention.

But the longer I live, the less convinced I become that ownership tells the whole story.

The deed may have your name on it.

The robin has never seen the paperwork.

The squirrels remain unimpressed.

The fungi beneath the soil continue conducting their ancient business without consulting county records.

Rain falls wherever gravity suggests.

Wind ignores fences.

Roots occasionally wander.

Seeds certainly do.

Nature acknowledges very few property lines.

Life has always preferred watersheds to lot lines.

This is not an argument against ownership.

It is an invitation to think beyond it.

The great conservationist Aldo Leopold proposed what he called a land ethic–the idea that humans are members of the ecological community rather than conquerors standing above it.

When I first encountered that idea years ago, I admired it intellectually.

Now, I find myself appreciating it personally.

Because membership changes behavior.

Owners ask,

"What are my rights?"

Members also ask,

"What are my responsibilities?"

That second question quietly changes everything.

One of my neighbors inherited a magnificent sugar maple from the family who owned the house before them.

He didn't plant it.

He simply became its next caretaker.

Every autumn he rakes leaves.

Every few years an arborist trims dead limbs.

During dry summers he waters it.

He often says,

"I'm just trying to keep it healthy for whoever lives here after us."

I love that sentence.

It contains no mention of resale value.

No calculation of return on investment.

Just stewardship.

Perhaps all meaningful ownership eventually matures into stewardship.

Parents understand this.

Teachers do too.

We do not own children.

We guide them for a while.

Communities borrow schools from previous generations and improve them for the next.

Libraries contain books written by people long gone, waiting patiently for readers not yet born.

Good civilizations think this way.

They inherit.

They improve.

They pass along.

The Earth deserves the same courtesy.

There is a Native American proverb often paraphrased as saying that we do not inherit the Earth from our ancestors; we borrow it from our children. Whether or not those exact words originate where they are often attributed, the wisdom they express has echoed across many cultures for generations.

Borrowing changes the conversation.

When you borrow a friend's canoe, you return it clean.

When you borrow a library book, you don't underline every page with permanent marker.

When someone lends you a favorite tool, you treat it more carefully than if it belonged to you.

Borrowing naturally invites gratitude.

Ownership occasionally encourages entitlement.

That distinction has become increasingly important to me.

Several years ago, I planted a young oak tree.

It stood no taller than my shoulder.

I watered it faithfully.

Protected it from deer.

Mulched around its base.

Each spring I measured its growth with unreasonable optimism.

One day I realized something amusing.

The tree would likely outlive me.

By a considerable margin.

Eventually someone else will mow around it.

Someone else will watch birds nest in its branches.

Children I will never meet may one day read beneath its shade.

I smiled.

The tree had quietly become a gift addressed to strangers.

There are worse investments.

Nature specializes in long-term thinking.

Acorns are perhaps the finest example.

An oak produces thousands of them.

Most never become trees.

Some feed squirrels.

Some nourish deer.

Some enrich the soil.

A few find favorable conditions.

Fewer still become towering oaks.

Nature seems remarkably comfortable with generous effort producing uncertain outcomes.

Human beings often struggle with this.

We like guarantees.

Nature prefers possibilities.

That has become one of my favorite lessons.

Plant the tree.

Restore the stream.

Teach the child.

Protect the meadow.

You may never witness the full harvest.

Do the work anyway.

The world is filled with blessings created by people who never lived to enjoy them.

Someone planted the trees shading your favorite park.

Someone preserved the trail you now hike.

Someone decided not to build on a piece of land because they believed beauty had value beyond immediate profit.

Someone restored a wetland.

Someone protected an eagle's nest.

Someone planted milkweed.

Someone cared.

You and I are beneficiaries of invisible generosity.

Perhaps the finest way to express gratitude is to become invisible benefactors ourselves.

There is another lesson hidden in shade.

Trees never enjoy their own.

The branches producing cool relief stand fully exposed to the sun.

Every act of generosity resembles that.

Parents lose sleep so children can rest.

Teachers prepare lessons long before students arrive.

Volunteers quietly stack chairs after community events.

Forests produce shade for creatures they will never meet.

Giving has always involved standing in the sunlight so someone else can sit comfortably beneath the leaves.

One afternoon, I watched a young boy reading beneath a large sycamore in the town park.

He looked perfectly at home.

Occasionally he glanced upward when a breeze stirred the leaves.

I wondered whether he realized the tree was older than his grandparents.

Probably not.

It didn't matter.

The tree wasn't asking to be admired.

It was simply doing what trees have always done.

Offering more than they receive.

That strikes me as an admirable ambition for human beings as well.

The older I become, the less interested I am in leaving monuments.

Stone eventually erodes.

Bronze tarnishes.

Buildings become outdated.

I would rather leave living things.

Trees.

Gardens.

Restored creeks.

Children who know the names of birds.

Communities where butterflies still drift through summer afternoons.

Places where someone can sit in unexpected shade and think kind thoughts about a person they never knew.

That seems enough.

Perhaps more than enough.

The storm eventually removed the old maple from our neighborhood.

Its trunk became lumber.

Its branches became mulch.

Its roots slowly returned to the soil.

Nothing was wasted.

The following spring, dozens of tiny maple seedlings appeared nearby.

Helicopter seeds, carried by the wind the previous autumn, had quietly begun another generation.

Standing there, I was reminded that nature rarely writes endings.

It writes continuations.

The old tree was gone.

The story wasn't.

Maybe that is the deepest truth about borrowing a planet.

Our task has never been to possess it.

Our task has always been to add one thoughtful chapter before handing the story to someone else.

The shade tree was never really ours.

We were simply fortunate enough to spend a little while beneath its branches.

And if we have borrowed well, someone we will never meet will someday stand in the cool afternoon beneath another tree, grateful for a kindness whose author they will never know.

Chapter Seventeen

The Long Conversation Between Generations

One autumn afternoon, I found myself standing in the oldest cemetery in our town.

Now, before you begin worrying about my social calendar, let me explain.

I enjoy old cemeteries.

Not in a gloomy sort of way.

Quite the opposite.

They are among the most hopeful places I know.

Old cemeteries are remarkably quiet.

No one is trying to impress anyone.

No advertisements.

No deadlines.

No debates about whose lawn looks better.

Just names.

Dates.

And the humbling reminder that every person buried there once believed tomorrow would arrive exactly as planned.

It rarely does.

As I wandered among weathered stones, I noticed something unexpected.

Several enormous maple trees shaded the paths.

Judging from their size, they had been planted well over a century earlier.

Whoever planted them never sat beneath the shade they would one day provide.

They planted anyway.

That simple realization has stayed with me.

The older I become, the more convinced I am that civilization is built almost entirely upon acts of generosity directed toward strangers.

Someone built the road you drove this morning.

Someone established the library where you discovered a favorite book.

Someone preserved the hiking trail where your family took photographs.

Someone planted trees beneath which generations of children would someday eat ice cream, read novels, and fall in love.

Most of those people never met the beneficiaries of their kindness.

They acted anyway.

The Earth has always depended on this kind of thinking.

Forests certainly do.

An old oak produces acorns it will never see mature.

Wildflowers scatter seeds without knowing where they will land.

Salmon fight their way upstream to reproduce before their lives come to an end, nourishing forests with the nutrients they carry from the sea.

Nature seems wonderfully comfortable investing in futures it will never personally witness.

Human beings occasionally forget this.

We have become remarkably skilled at measuring quarters, fiscal years, election cycles, and annual reports.

Nature measures generations.

The difference matters.

A forest does not panic because a sapling is only three feet tall.

The forest understands the mathematics of patience.

I confess that I have not always shared its confidence.

Like many people, I have occasionally planted tomatoes and then inspected them the following morning with unreasonable expectations.

"Really?" I mutter.

"Nothing yet?"

The tomatoes remain politely unconcerned.

Growth has its own calendar.

The same is true of children.

Communities.

Friendships.

Ideas.

Justice.

Hope.

The finest things in life generally refuse to hurry.

Perhaps that is why older people sometimes smile when younger people become impatient.

We've watched enough seasons to know that many things cannot be rushed.

The first daffodil does not bloom because we stare at it.

It blooms because spring quietly arrives.

One of the great privileges of teaching for many years is that former students occasionally return.

Sometimes decades later.

Gray hair appears where backpacks once hung.

Children now accompany people who once asked whether assignments would be on the exam.

The conversations are always fascinating.

Rarely do they remember my lectures.

Thankfully.

Some of those PowerPoint presentations deserved to be forgotten.

Instead, they remember conversations after class.

An encouraging comment.

A book recommendation.

A story.

A joke.

One student reminded me of something I had said nearly thirty years earlier.

I had no recollection of it.

She had built part of her career around it.

Teaching has a way of revealing an important truth.

You almost never know which seeds are growing.

The same is true of parenting.

Mentoring.

Volunteering.

Neighboring.

Planting trees.

Life is wonderfully inefficient.

You scatter encouragement widely and trust that something

beautiful will eventually emerge.

Nature has never demanded certainty before allowing generosity.

Several years ago, I visited a restored prairie.

A century earlier it had been farmland.

Later it became neglected ground.

Eventually volunteers began patiently reintroducing native grasses and wildflowers.

The first few years looked discouraging.

Weeds dominated.

Many seedlings failed.

Visitors questioned whether the project would ever succeed.

Today, the prairie hums with life.

Butterflies drift above coneflowers.

Meadowlarks sing.

Native bees move from blossom to blossom.

Children chase grasshoppers through grasses taller than they are.

The restoration succeeded because people were willing to work for a landscape they might never fully experience.

Hope always contains an element of delayed gratification.

One of my favorite words in the English language is legacy.

Unfortunately, it has acquired a somewhat glamorous reputation.

People imagine buildings named after donors.

Bronze statues.

Portraits.

Endowed chairs.

Those things have their place.

Nature offers a quieter definition.

Legacy is fertile soil.

Clean water.

Healthy forests.

Children who know the names of birds.

Communities where pollinators still thrive.

Rivers where salmon return.

Neighborhoods where old trees remain standing because someone decades earlier decided they mattered.

Most meaningful legacies are anonymous.

You have benefited from thousands of them already.

Someone protected the watershed supplying your drinking water.

Someone decided not to pave over a wetland.

Someone preserved open space where deer still wander at dawn.

Someone insisted that a magnificent old tree should remain standing despite easier alternatives.

You probably don't know their names.

You enjoy their gifts every day.

That realization has changed the way I think about gratitude.

Gratitude should not only travel backward.

It should travel forward.

The best way to thank previous generations is to become good ancestors ourselves.

I love that phrase.

A good ancestor.

It has nothing to do with age.

You can become a good ancestor at twenty-five.

Or fifteen.

Or seventy-five.

The qualification is surprisingly simple.

Make decisions that increase the chances of future flourishing.

Plant trees.

Restore habitats.

Reduce waste.

Teach curiosity.

Protect beauty.

Encourage kindness.

Leave places healthier than you found them.

The Earth recognizes all of these as investments.

One evening I watched a grandfather teaching his grandson how to plant an oak.

The boy held the sapling while the older man filled the hole with soil.

When they finished, the child asked,

"When will it get really big?"

The grandfather smiled.

"Probably after I'm gone."

The boy looked puzzled.

"Then why are we planting it?"

The old man leaned on his shovel for a moment.

"So you'll have somewhere to bring your grandchildren."

That may be the finest explanation of stewardship I have ever heard.

Borrowing a planet has never been about maximizing what we can take during our brief visit.

It has always been about enlarging the possibilities available to those who arrive after us.

Every generation receives an unfinished world.

Some inherit cleaner rivers.

Others inherit damaged forests.

Some receive healthy soil.

Others receive depleted fields.

Every generation has the same choice.

Leave the conversation richer than you found it.

The remarkable thing is that nature forgives slow progress.

Plant one tree.

Teach one child.

Restore one stream.

Protect one meadow.

Encourage one neighbor.

The forest never asks a single acorn to become the entire woodland.

It simply asks it to become an oak.

The rest follows.

As I left the cemetery that autumn afternoon, I paused beneath one of those magnificent old maples.

Its branches stretched high above the path, scattering golden leaves in the afternoon sunlight.

Someone I would never meet had planted it.

Someone whose name had probably long since faded from memory.

Yet their kindness continued.

I stood in their shade.

The thought made me smile.

Perhaps immortality has always looked less like being remembered and more like creating goodness that continues after your name is forgotten.

That possibility strikes me as wonderfully freeing.

We need not become famous.

We simply need to become faithful.

Plant the tree.

Teach the child.

Leave the porch light on.

Borrow this remarkable planet with gratitude.

Then hand it forward just a little healthier, a little kinder, and a little more beautiful than the way you found it.

That is how generations continue their conversation.

One thoughtful act at a time.

Chapter Eighteen

Enough Is a Beautiful Word

Several years ago, I bought a birdhouse.

Not because I needed one.

Because it was on sale.

Now, I have reached the stage of life where sales possess an almost mystical power.

I can resist temptation.

I can ignore advertising.

I can walk confidently past entire departments filled with things I neither need nor understand.

But place a small sign that says 30% Off, and suddenly I begin inventing entirely fictional scenarios in which a decorative garden gnome becomes an essential household investment.

The birdhouse was handsome.

Cedar.

Copper roof.

Excellent craftsmanship.

I brought it home feeling quietly pleased with myself.

Then I walked into the backyard.

There, hanging from the maple tree, was another birdhouse.

One I had completely forgotten I already owned.

It was occupied.

The wrens, apparently unaware of current retail pricing, had selected last year's model.

I stood there holding the new birdhouse and laughed.

Not because I had wasted money.

Although there was that.

I laughed because the birds had just reminded me of a lesson I seem to need every few years.

Enough had already arrived.

I simply hadn't noticed.

There are words that become more beautiful with age.

Patience.

Neighbor.

Gratitude.

Home.

Another of those words is enough.

Children rarely understand it.

Young professionals often postpone it.

Advertisers certainly avoid it.

Nature, however, seems entirely comfortable with the idea.

An oak tree does not attempt to become the tallest organism on Earth.

A chickadee gathers enough food for winter.

A fox hunts until hunger is satisfied.

Wildflowers bloom as extravagantly as they can, yet they never apologize when autumn asks them to rest.

The natural world possesses an astonishing confidence in sufficiency.

Human beings often struggle with that confidence.

Our culture has become remarkably skilled at whispering a single sentence into our ears.

A little more.

A little more money.

A little larger house.

A little newer car.

A few more followers.

A few more accomplishments.

A little more recognition.

The finish line has an unfortunate habit of relocating just before we reach it.

The philosopher in me finds this fascinating.

The gardener in me finds it exhausting.

One spring I planted tomatoes.

By midsummer they produced more tomatoes than any reasonable household could possibly consume.

Neighbors received tomatoes.

Friends received tomatoes.

Complete strangers slowed their cars because they had heard rumors.

Even after making sauce, salsa, and enough tomato sandwiches to qualify as a seasonal lifestyle, the plants continued producing.

The tomatoes had never read an economics textbook.

They weren't maximizing profit.

They were simply participating in abundance.

Nature distinguishes between abundance and accumulation.

Abundance is generous.

Accumulation is anxious.

Abundance says,

"There is enough to share."

Accumulation quietly worries,

"Perhaps there won't be."

The distinction matters.

Walk through an old forest.

Nothing appears hurried.

The trees are not racing.

The ferns are not competing for magazine covers.

The stream is not trying to become a waterfall.

Everything is fully engaged in becoming itself.

Perhaps that is enough.

This is not an argument against ambition.

Please don't misunderstand me.

Ambition built hospitals.

Sent people to the moon.

Eradicated diseases.

Created libraries.

Composed symphonies.

Restored ecosystems.

Human progress owes much to people unwilling to accept unnecessary limits.

But there are two kinds of ambition.

One grows from gratitude.

The other grows from scarcity.

Grateful ambition asks,

"How can I contribute?"

Scarcity asks,

"When will I finally have enough?"

Only one of those questions has a satisfying answer.

I remember visiting the home of an elderly couple who had lived there for nearly sixty years.

The house wasn't large.

The furniture showed the honest wear of long companionship.

Books occupied every available shelf.

Photographs filled the walls.

Outside stood an enormous sugar maple shading a vegetable garden.

As we sat on the porch drinking lemonade, I asked whether they had ever considered moving into something larger.

The husband looked around the yard for a long moment.

Then he smiled.

"We've spent sixty years making this place fit our lives."

He paused.

"Now our lives fit this place."

I have carried that sentence with me ever since.

Perhaps contentment is not settling.

Perhaps it is finally noticing.

The environmental conversation often focuses on technology.

Renewable energy.

Efficient transportation.

Cleaner manufacturing.

All of these matter enormously.

Human ingenuity will continue playing an essential role in caring for our planet.

But I sometimes wonder whether one of the greatest environmental technologies has existed all along.

The ability to say,

"This is enough."

Enough food without waste.

Enough possessions without excess.

Enough success without sacrificing every relationship.

Enough work to support a meaningful life.

Enough consumption to meet genuine needs while leaving room for other living things to flourish.

Enough.

It is a small word carrying remarkable freedom.

The ecologists remind us that every ecosystem has limits.

A pond supports only so many fish.

A forest only so many deer.

A meadow only so many nesting birds.

Those limits are not punishments.

They are the mathematics of balance.

Nature rarely asks,

"How much can I take?"

It quietly demonstrates,

"How much allows everything to thrive?"

There is wisdom there.

Not only for ecosystems.

For people.

One autumn afternoon I watched squirrels gathering acorns beneath an old oak.

Each squirrel buried acorns in dozens of different places.

Not all in one location.

Not endlessly collecting.

Just enough.

Some would become winter meals.

Some would be forgotten.

Those forgotten acorns would eventually become young oaks.

The squirrel thought it was preparing for winter.

The forest knew it was planting tomorrow.

One action.

Two purposes.

Nature delights in elegant solutions.

The older I become, the less impressed I am by lives that continually become larger.

I find myself admiring lives that become deeper.

People who know their neighbors.

Who remember birthdays.

Who grow tomatoes and give most of them away.

Who stop to help turtles cross roads.

Who plant milkweed for butterflies they may never personally see.

Who leave campsites cleaner than they found them.

Who quietly improve the small corner of Earth entrusted to them.

None of this attracts much attention.

It attracts something better.

Peace.

There is another reason I treasure the word enough.

It leaves room.

When every shelf is full, nothing new can be added.

When every hour is scheduled, wonder has nowhere to land.

When every landscape is developed, birds lose places to nest.

When every conversation is about ourselves, we stop hearing other voices.

Enough creates breathing room.

For people.

For forests.

For rivers.

For silence.

For gratitude.

Perhaps that is why Sabbath appears in so many cultures.

Not merely as religious practice.

As ecological wisdom.

A reminder that endless production eventually exhausts both the land and the soul.

Even fields once rested.

Even forests have seasons of apparent stillness.

Creation itself seems to understand the necessity of pause.

One evening, as the sun disappeared behind the trees, I noticed the old birdhouse hanging from the maple.

The wrens had long since finished raising their young.

The entrance hole sat empty.

The birdhouse looked weathered now.

Its paint had faded.

A corner had begun to split.

I realized it was beautiful precisely because it had been useful.

The new birdhouse still waited in the garage.

Perfect.

Unused.

Perhaps that is another lesson.

The purpose of life is not remaining pristine.

It is becoming well used in the service of something worthwhile.

Enough.

It is not a word of resignation.

It is a word of liberation.

Enough reminds us that joy is often discovered before the next purchase.

That gratitude frequently arrives before the next achievement.

That stewardship begins the moment we stop asking how much more we can acquire and start asking how generously we can care for what we have already been given.

The Earth has never asked us to possess everything.

It has simply invited us to cherish enough.

There is extraordinary beauty in that invitation.

More, I have discovered, than I ever found in another birdhouse on sale.

Chapter Nineteen

Leaving the Campsite Better Than You Found It

One of the finest pieces of advice I ever received came from someone carrying a frying pan.

That alone should remind us never to underestimate camp cooks.

Years ago, while camping with a group of friends, I watched an older gentleman finish breakfast.

He washed the pan.

Collected stray napkins.

Picked up a candy wrapper that none of us had noticed.

Brushed leaves back over a patch of flattened ground.

Then he walked a few yards into the woods, returned with two aluminum cans left by someone else, and tucked them into the recycling bag.

I finally asked him,

"Did someone ask you to do all that?"

He looked genuinely puzzled.

"No."

"Then why?"

He smiled.

"Because the next person deserves to think this place has always been beautiful."

That sentence has followed me for years.

Not just on camping trips.

Into classrooms.

Meetings.

Neighborhoods.

Friendships.

Gardens.

Libraries.

Conversations.

It has become one of those ideas that quietly expands every time you look at it.

Leave the campsite better than you found it.

Simple.

Practical.

Almost embarrassingly obvious.

And surprisingly rare.

We tend to think of stewardship as something grand.

Protecting endangered species.

Restoring wetlands.

Passing important legislation.

Funding scientific research.

Those efforts matter tremendously.

But stewardship also lives in much smaller moments.

Returning the shopping cart.

Picking up litter that isn't yours.

Planting one tree.

Teaching a child the name of a bird.

Repairing a park bench.

Refilling a bird feeder.

Stopping to help a turtle cross the road.

None of these actions will appear in history books.

History, fortunately, is not the only place goodness accumulates.

Nature has always favored accumulation.

One leaf becomes soil.

One acorn becomes a forest.

One raindrop joins a river.

One bee visits one flower.

Again.

And again.

Until orchards bloom and harvests become possible.

The world rarely changes all at once.

It changes through countless ordinary acts repeated faithfully.

That realization has made me suspicious of dramatic gestures.

Not because they lack value.

Because they sometimes distract us from quieter opportunities.

Most environmental damage happens one decision at a time.

So does most environmental healing.

Several years ago, a neighborhood near ours organized a Saturday morning cleanup along a small creek.

Nothing glamorous.

No television cameras.

No famous speakers.

Just people wearing work gloves and carrying trash bags.

Old tires emerged from the water.

Plastic bottles.

Shopping carts.

Enough golf balls to suggest either remarkable optimism or remarkably poor aim.

By lunchtime, the creek looked transformed.

A week later, something unexpected happened.

People began walking there.

Children explored the banks.

Birdwatchers appeared with binoculars.

Someone built a small bench overlooking the water.

Within a year, neighbors who had driven past the creek for decades suddenly spoke of our creek.

The cleanup had removed more than litter.

It had restored the relationship.

Stewardship often begins by changing pronouns.

From the park...

to our park.

From the river...

to our river.

From that neighborhood...

to our neighborhood.

Ownership isn't the issue.

Belonging is.

One of the remarkable things about belonging is that it changes what we notice.

Have you ever observed someone cleaning a house before guests arrive?

Dust that remained invisible for weeks suddenly becomes astonishingly obvious.

Crumbs develop magnetic properties.

Cobwebs appear to have multiplied overnight.

Love sharpens attention.

The same thing happens when we begin thinking of the Earth as home instead of scenery.

Litter becomes harder to ignore.

Healthy trees become easier to appreciate.

A stream becomes more than drainage.

A vacant lot becomes a possibility.

Once affection arrives, indifference becomes surprisingly difficult.

There is another lesson hidden in campsites.

You never really camp alone.

Even when no one else is nearby, someone is there before you.

Someone will arrive after you.

Every campsite is part of a long conversation among strangers who never meet.

One family gathers firewood.

Another leaves extra kindling.

Someone repairs a picnic table.

Someone forgets to extinguish a fire completely.

Someone picks up trash left by another camper.

Every visitor inherits the consequences of previous visitors.

Sound familiar?

It should.

That's civilization.

The same is true of the planet.

Every generation receives rivers shaped by earlier decisions.

Forests protected–or cleared–by previous hands.

Air cleaner or dirtier, depending on choices already made.

Healthy soil or exhausted fields.

We arrive in the middle of a story.

We never begin it.

We never finish it.

Our responsibility is wonderfully modest.

Improve the next chapter.

Do not write the entire book.

One chapter.

That's enough.

The older I become, the more I appreciate modest ambitions.

Plant one oak.

Restore one stream.

Mentor one student.

Support one community garden.

Reduce one source of waste.

Encourage one neighbor.

Leave one place a little healthier than you found it.

These actions rarely feel dramatic.

Forests understand why.

An individual tree does not create a woodland.

It simply grows.

Thousands of similar acts eventually become a forest.

I once visited a botanical garden where volunteers had spent decades restoring native woodland.

One woman had been working there every Thursday morning for nearly twenty-five years.

I asked what she had accomplished.

She looked around.

"Oh," she said, "mostly weeding."

That answer delighted me.

Mostly weeding.

No heroic speeches.

No monuments.

No headlines.

Just twenty-five years of removing obstacles so other lives could flourish.

The garden was magnificent.

Its beauty rested upon thousands of ordinary Thursdays.

Perhaps the world is repaired that way.

One ordinary Thursday at a time.

One of my favorite questions has become remarkably simple.

When I leave a place, I ask myself,

"Is this place even slightly better because I was here?"

Not dramatically better.

Slightly.

Did someone smile?

Did I learn a bird's name?

Did I encourage another person?

Did I leave less waste behind?

Did I notice beauty?

Did I create any?

Those questions are refreshingly achievable.

They replace anxiety with opportunity.

The Earth does not ask us to solve every problem before dinner.

It asks us to participate.

There is enormous freedom in that.

Late one afternoon, while hiking a woodland trail, I noticed a young family stopped beside a fallen branch blocking the path.

The father quietly dragged it aside.

The children helped move smaller sticks.

It took perhaps two minutes.

Then they continued walking.

The family behind them never knew the branch had been there.

That, I realized, is stewardship at its finest.

Someone else enjoys an easier journey because you quietly paused.

No applause. No recognition.

Just kindness made practical.

Nature itself seems to operate on that principle.

Trees enrich the soil they will never use.

Flowers feed pollinators they will never meet again.

Rivers nourish communities downstream.

Clouds rain on fields belonging to strangers.

The Earth has always been left better than it was found.

Perhaps we have been invited to imitate its generosity.

When my camping friend finished cleaning the site that morning, he took one last look around.

Nothing appeared different.

No visitor would have noticed the extra effort.

Except they would.

Not consciously.

They would arrive in a place that felt cared for.

I've come to believe that human beings can sense stewardship.

We notice when a garden has been loved.

When a trail has been maintained.

When a neighborhood has been cherished.

Care leaves fingerprints.

Usually invisible.

Always real.

Eventually, every one of us will leave this remarkable campsite we call Earth.

The tents will come down.

The fire will grow cold.

Someone else will unpack beneath the same stars.

I hope they find cleaner streams.

Healthier forests. More birds.

Kinder neighborhoods.

Children who still know how to wonder.

And perhaps, without ever knowing our names, they will quietly think,

"Whoever was here before us cared about this place."

If that happens, I suspect we will have borrowed the planet well.

Chapter Twenty

One More Walk Around the Block

There is a route I have walked hundreds of times.

Perhaps thousands.

It begins at my front door.

Turns left at the corner.

Passes the old maple.

Crosses a former railroad track.

Circles the neighborhood park.

Then wanders home again.

The entire walk takes about thirty minutes.

Forty-five if I stop to talk with neighbors.

An hour if I encounter children with questions about turtles.

Slightly longer if squirrels are behaving unusually.

Which, in fairness, is often the case.

Nothing about the route would impress a travel writer.

There are no waterfalls.

No mountain vistas.

No ancient castles.

No dramatic overlooks where people stand in expensive hiking boots while inspirational music quietly swells in the background.

It is simply my neighborhood.

Ordinary.

Familiar.

Home.

Or at least, that's what I used to think.

These days, I know better.

Because over the years,s the walk has quietly changed.

Not because the neighborhood became more interesting.

Because I did.

The maple is no longer just a tree.

I know which branches the chickadees prefer.

I know where the woodpecker usually begins his morning inspection.

Every October, I look for the first leaves to turn gold.

Every April,l I watch the buds swell just enough to hint that winter is finally surrendering.

We've become acquaintances.

The creek beneath the bridge no longer appears to be simply water moving from one place to another.

After heavy rains, it speaks loudly.

During summer droughts, it whispers.

I know where the minnows gather.

Where the turtles climb onto old logs to warm themselves.

Where the kingfisher occasionally waits with remarkable patience before plunging into the current.

The creek has introduced itself.

The park has done the same.

The old sycamore whose bark peels like camouflage.

The bench that receives the evening sun.

The patch of milkweed where monarchs pause in late summer.

The meadow that hums with bees in June.

The owls whose conversations begin just after sunset.

None of these things suddenly appeared.

They had been waiting for me.

Attention has a curious effect on time.

The more we notice, the more slowly life seems to unfold.

Children understand this instinctively.

An hour outside can become an expedition.

Adults often experience the opposite.

Entire weeks disappear without leaving much of a memory.

I have begun to suspect that memorable days are not necessarily the busiest ones.

They are the most attentive.

The days we remember are filled with particular things.

The first hummingbird of spring.

The smell of lilacs.

A squirrel crossing the road at dawn.

Fresh snow settling on pine branches.

A child proudly identifies a goldfinch without help.

Specific memories.

Not generalized busyness.

Attention leaves footprints in memory.

Distraction rarely does.

One evening not long ago, I found myself sitting on the familiar bench beneath the sycamore.

The sun was setting toward the horizon.

Children were finishing one last game before dinner.

A dog enthusiastically retrieved a tennis ball with the conviction that no previous retrieval had ever been quite so successful.

A pair of geese flew overhead, announcing themselves to everyone within three counties.

Somewhere nearby, someone was grilling hamburgers.

The smell drifted through the warm air.

A breeze carried the fragrance of freshly cut grass.

Nothing extraordinary happened.

Which is precisely the point.

The extraordinary had been quietly disguised as ordinary all along.

A little girl stopped beside the bench.

She couldn't have been more than six years old.

She held something carefully cupped in her hands.

"Mister," she whispered.

"I found a snail."

She opened her hands just enough for me to see.

The snail, apparently unconcerned about becoming the center of attention, continued its slow journey across her palm.

"Isn't he amazing?" she asked.

Now, there are several ways an adult can answer that question.

You can provide information.

You can explain that snails carry their homes with them.

You can describe mollusks, shells, calcium, locomotion, and moisture.

All of those answers are accurate.

Instead, I smiled.

"He certainly is."

She nodded, completely satisfied.

Then she carried the snail toward a patch of damp grass, where she gently placed him beneath a fern.

Off he went.

Neither of us applauded.

Nature rarely requires applause.

As she skipped away, I realized that she had understood the entire message of this book without ever reading a single chapter.

She noticed.

She admired.

She cared.

The sequence was effortless.

Curiosity became wonder.

Wonder became kindness.

Perhaps stewardship has always been that simple.

Not easy.

Simple.

The Earth has never asked us to memorize every bird.

To identify every wildflower.

To understand every ecological relationship.

It has simply invited us into one long conversation.

Learn a few names.

Ask a few questions.

Plant a tree.

Protect a stream.

Share a trail.

Watch the clouds.

Leave the campsite a little better than you found it.

Then repeat.

One walk at a time.

One season at a time.

One generation at a time.

When I was younger, I believed meaningful lives were built from extraordinary moments.

Milestones.

Achievements.

Big decisions.

Important destinations.

Those things matter.

But I have come to suspect that a meaningful life is actually assembled from thousands of ordinary Tuesdays that receive our full attention.

A cup of coffee on the porch while chickadees argue over sunflower seeds.

A grandchild introducing you to a caterpillar you've walked past a hundred times.

The first tomato is ripening in the garden.

The last firefly of summer.

The first snowfall settling quietly over sleeping trees.

A familiar walk around the block that slowly becomes a pilgrimage.

We spend much of our lives searching for somewhere else.

A better view.

A larger adventure.

A more extraordinary landscape.

All the while, the remarkable world beneath our feet waits patiently for us to notice that we have already arrived.

Perhaps that has always been the secret.

Wonder is not hiding.

It is waiting.

Gratitude is not manufactured.

It grows wherever attention is planted.

Stewardship is not born from guilt.

It grows naturally from affection.

And affection begins with noticing.

I have walked this neighborhood for many years now.

Someday another family will live in this house.

Another pair of feet will cross the little bridge.

Another child will discover turtles in the creek.

Another gardener will complain about squirrels with entirely justified exasperation.

The maple will continue growing.

The robins will continue singing.

Spring will continue arriving with astonishing confidence.

The Earth will continue extending invitations.

I hope they accept them.

I hope they learn the names of their neighbors, both human and wild.

I hope they plant something whose shade they will never enjoy.

I hope they leave the porch lights on.

I hope they leave the campsite just a little better than they found it.

Most of all, I hope they never become too busy to stop for a snail crossing the path.

Because the measure of a life has never been how far we traveled.

It has been how deeply we belonged to the places that welcomed us.

The planet was never ours to own.

It was ours to borrow.

To notice.

To cherish.

To care for.

To hand forward with gratitude.

Tomorrow morning, the sun will rise again.

The birds will begin their ancient songs before most alarm clocks have the courtesy to ring.

The trees will quietly continue making oxygen.

The fungi will continue enriching the soil.

The bees will leave for work.

The clouds will drift wherever the wind invites them.

The Earth will once again perform countless miracles without charging admission.

Go outside.

Take one more walk around the block.

This time, introduce yourself.

You've been living among neighbors all along.

Epilogue

Before You Return the Keys

If you've made it this far, I have a confession.

This was never really a book about the environment.

Not entirely.

It was a book about a relationship.

About the quiet miracle that something changes inside us the moment we stop asking,

"What can I get from this world?"

and begin asking,

"How should I care for it while I'm here?"

That question has a curious way of spilling into every corner of life.

Suddenly, we become more patient with children.

Kinder to strangers.

More forgiving of ourselves.

More grateful for ordinary Tuesdays.

We notice old trees.

We remember our neighbors' names.

We begin planting things we may never live to see fully grown.

The world itself hasn't changed.

Our posture toward it has.

For years, I believed wisdom was mostly about collecting answers.

Age has politely corrected me.

Wisdom seems much more interested in learning to pay attention.

Attention to beauty.

Attention to suffering.

Attention to birdsong before breakfast.

Attention to the quiet loneliness of another human being.

Attention to a spiderweb glistening with dew that yesterday I would have walked past without a glance.

Attention is the beginning of love.

And love, sooner or later, always becomes stewardship.

One of the unexpected gifts of growing older is discovering that life becomes simultaneously smaller and larger.

Smaller because your world gradually gathers around familiar places.

A favorite chair.

A neighborhood walk.

A porch.

A garden.

The old maple at the corner.

The coffee mug with the chipped handle that somehow pours optimism every morning.

Larger because you begin seeing those ordinary things as part of an astonishingly connected whole.

The rain falling on your roof also nourishes distant forests.

The bees visiting your flowers are part of an ancient partnership older than civilization.

The child asking about a caterpillar is carrying forward questions that have inspired every scientist, poet, and philosopher who ever lived.

Everything belongs.

Including you.

Especially you.

The Earth has always been remarkably generous with its introductions.

Every sunrise says,

"Good morning."

Every spring says,

"Let's try again."

Every autumn says,

"Learning to let go can also be beautiful."

Every snowfall whispers,

"There is still time to begin fresh."

The lessons have always been available.

The classroom has never closed.

Admission has always been free.

If there is one hope I have for you after closing this book, it is not that you become an expert naturalist.

I hope you learn a few more names.

The tree outside your window.

The bird that visits every morning.

The wildflowers are blooming beside the road.

The neighbor you've waved to for years but never quite met.

Names have a way of turning strangers into neighbors.

And neighbors are much harder to ignore.

You don't have to save the entire planet.

No one person ever could.

The Earth has never asked that of us.

It asks something far more hopeful.

Care faithfully for the little corner entrusted to you.

Your porch.

Your classroom.

Your stream.

Your family.

Your community garden.

Your conversations.

Your ordinary Tuesdays.

That has always been enough.

One thoughtful life has a surprising reach.

Long after you've forgotten a conversation, someone may still remember your encouragement.

Long after you've planted a tree, someone may rest beneath its branches.

Long after you've picked up a piece of litter, a child may assume that clean places are simply the way the world is supposed to look.

Kindness echoes.

So does stewardship.

Perhaps they are the same thing wearing different clothes.

Someday, many years from now, I hope you and I will each return the keys to this borrowed planet.

Others will walk these same trails.

They will listen to the descendants of today's birds.

They will sit beneath trees we never saw reach their full height.

They will inherit streams that are cleaner or dirtier because of the choices we made.

They will never know most of our names.

That is perfectly all right.

The forest has never insisted on receiving credit.

Neither do gardens.

Nor rivers.

Nor grandparents.

The finest gifts often arrive anonymously.

When that day comes, I hope the world feels gently used rather than heavily consumed.

I hope there are more butterflies.

More old trees.

More children who know how to kneel beside a snail without checking their phones.

More neighbors who leave porch lights on.

More people who understand that enough is a beautiful word.

More lives measured not by what they accumulated, but by what they quietly helped flourish.

And if, somewhere in the years ahead, you happen to be walking around your neighborhood on an entirely ordinary evening...

If you notice a bird you've never seen before...

Or stop to admire the first crocus pushing through the last snow...

Or smile at the sound of geese returning in spring...

Or help a turtle cross the road...

Or plant a tree whose shade will belong to someone else...

Then perhaps you'll remember this little conversation we shared.

Not because these pages contained anything extraordinary.

But because they reminded you of something you already knew.

The world has never been ordinary.

It has always been alive with invitations.

Invitations to notice.

To wonder.

To laugh.

To belong.

To care.

To become, for a little while, a gracious borrower.

So before you put this book back on the shelf, do one small thing for me.

Step outside.

Leave your phone in your pocket.

Walk slowly.

Look up.

Then look down.

Listen longer than feels necessary.

Learn one new name.

Smile at one stranger.

Thank one tree.

Wave to one neighbor.

Stand quietly until the world introduces itself again.

It has been waiting for you.

Welcome home.

Bibliography

Abram, D. (1996). *The spell of the sensuous: Perception and language in a more-than-human world.* Vintage Books.

Attenborough, D. (2020). *A life on our planet: My witness statement and a vision for the future.* Grand Central Publishing.

Berry, T. (1999). *The great work: Our way into the future.* Bell Tower.

Berry, W. (1981). *The gift of good land: Further essays cultural and agricultural.* North Point Press.

Berry, W. (2002). *The art of the commonplace: The agrarian essays.* Counterpoint.

Bohm, D. (1980). *Wholeness and the implicate order.* Routledge.

Brazelton, T. B., & Greenspan, S. I. (2000). *The irreducible needs of children.* Perseus Publishing.

Brooks, D. (2023). *How to know a person: The art of seeing others deeply and being deeply seen.* Random House.

Brown, B. (2010). *The gifts of imperfection.* Hazelden.

Brown, B. (2012). *Daring greatly.* Gotham Books.

Bryson, B. (2019). *The body: A guide for occupants.* Doubleday.

Bryson, B. (2020). *The nature of nature.* Doubleday.

Cajete, G. (1994). *Look to the mountain: An ecology of Indigenous education.* Kivaki Press.

Carson, R. (1951). *The sea around us.* Oxford University Press.

Carson, R. (1962). *Silent Spring.* Houghton Mifflin.

Carson, R. (1965). *The sense of wonder.* Harper & Row.

Csikszentmihalyi, M. (1990). *Flow: The psychology of optimal experience.* Harper & Row.

Darwin, C. (1859/2009). *On the origin of species.* Penguin Classics.

De Waal, F. (2016). *Are we smart enough to know how smart animals are?* W. W. Norton.

Diamond, J. (2005). *Collapse: How societies choose to fail or succeed.* Viking.

Dillard, A. (1974). *Pilgrim at Tinker Creek.* Harper's Magazine Press.

Dirzo, R., Young, H. S., Galetti, M., Ceballos, G., Isaac, N. J. B., & Collen, B. (2014). *Defaunation in the Anthropocene.* Science, 345(6195), 401–406.

Ehrlich, P. R., & Ehrlich, A. H. (1981). *Extinction.* Random House.

Eiseley, L. (1957). *The immense journey.* Vintage Books.

Eiseley, L. (1978). *The star thrower.* Harcourt Brace Jovanovich.

Frankl, V. E. (2006). *Man's search for meaning* (Rev. ed.). Beacon Press. (Original work published 1946)

Fuller, R. A., et al. (2007). *Psychological benefits of greenspace increase with biodiversity.* Biology Letters, 3(4), 390–394.

Gladwell, M. (2008). *Outliers.* Little, Brown and Company.

Goodall, J. (1999). *Reason for hope.* Grand Central Publishing.

Goodall, J. (2021). *The book of hope.* Celadon Books.

Harari, Y. N. (2015). *Sapiens: A Brief History of Humankind.* Harper.

Harris, S. (2014). *Waking up.* Simon & Schuster.

Hölldobler, B., & Wilson, E. O. (2009). *The superorganism.* W. W. Norton.

Kabat-Zinn, J. (1994). *Wherever you go, there you are.* Hyperion.

Keltner, D. (2023). *Awe: The new science of everyday wonder and how it can transform your life.* Penguin Press.

Kimmerer, R. W. (2013). *Braiding Sweetgrass: Indigenous Wisdom, Scientific Knowledge, and the Teachings of Plants.* Milkweed Editions.

Kimmerer, R. W. (2021). *Gathering moss.* Oregon State University Press.

Leopold, A. (1949). *A Sand County almanac.* Oxford University Press.

Louv, R. (2005). *Last child in the woods*. Algonquin Books.

Louv, R. (2011). *The nature principle.* Algonquin Books.

Macy, J., & Brown, M. Y. (2014). *Coming back to life.* New Society Publishers.

Macfarlane, R. (2015). *Landmarks.* Penguin Books.

Macfarlane, R. (2019). *Underland.* W. W. Norton.

Mayer, F. S., & Frantz, C. M. (2004). *The connectedness to nature scale*. Journal of Environmental Psychology, 24(4), 503–515.

McCullough, M. E. (2020). *The kindness of strangers.* Harper.

Meadows, D. H. (2008). *Thinking in systems.* Chelsea Green Publishing.

Muir, J. (1916). *A thousand-mile walk to the Gulf.* Houghton Mifflin.

Muir, J. (1997). *My first summer in the Sierra.* Dover Publications. (Original work published 1911)

National Academies of Sciences, Engineering, and Medicine. (2020). *Valuing climate damages.* National Academies Press.

Nisbet, E. K., Zelenski, J. M., & Murphy, S. A. (2009). *The Nature Relatedness Scale.* Environment and Behavior, 41(5), 715–740.

Odum, E. P., & Barrett, G. W. (2005). *Fundamentals of ecology (5th ed.)*. Brooks/Cole.

Orr, D. W. (1994). *Earth in mind.* Island Press.

Pinker, S. (2018). *Enlightenment now.* Viking.

Plotkin, B. (2008). *Nature and the human soul.* New World Library.

Popkin, G. (2023). *The matter of everything.* W. W. Norton.

Sagan, C. (1994). *Pale blue dot.* Random House.

Sagan, C. (1996). *The demon-haunted world.* Random House.

Schumacher, E. F. (1973). *Small is beautiful.* Harper & Row.

Seligman, M. E. P. (2011). *Flourish.* Free Press.

Snyder, G. (1990). *The practice of the wild.* North Point Press.

Suzuki, D. (1997). *The sacred balance.* Greystone Books.

Tallamy, D. W. (2007). *Bringing nature home.* Timber Press.

Tallamy, D. W. (2019). *Nature's best hope.* Timber Press.

Thoreau, H. D. (2004). *Walden.* Barnes & Noble Classics. (Original work published 1854)

Tuan, Y.-F. (1977). *Space and place: The perspective of experience.* University of Minnesota Press.

Ulrich, R. S. (1984). *View through a window may influence recovery from surgery.* Science, 224(4647), 420–421.

United Nations Environment Program. (2023). *Global environment outlook.* United Nations.

Wilson, E. O. (1984). *Biophilia.* Harvard University Press.

Wilson, E. O. (1992). *The diversity of life.* Harvard University Press.

Wilson, E. O. (2016). *Half-Earth: Our planet's fight for life.* Liveright.

Wohlleben, P. (2016). *The hidden life of trees.* Greystone Books.

World Wildlife Fund. (2022). *Living Planet* Report 2022. World Wildlife Fund.

Yalom, I. D. (2008). *Staring at the sun: Overcoming the terror of death.* Jossey-Bass.

About the Author

Mark Donnelly, PhD, has spent much of his life asking questions that begin with simple words like why, what if, and have you ever noticed...?

A former professor, marketing strategist, photographer, historian, and lifelong student of human behavior, he has spent more than three decades helping people think more clearly, look more closely, and rediscover the extraordinary hidden inside ordinary life. His work bridges psychology, philosophy, behavioral science, creativity, leadership, history, and the quiet wisdom found in everyday experiences.

He has authored more than sixty books exploring curiosity, happiness, wonder, resilience, critical thinking, creativity, and meaningful living. Across them all runs a common thread: the belief that the best questions are often more valuable than the quickest answers, and that a well-lived life begins with paying attention.

In *Instructions for Borrowing a Planet*, Donnelly brings together a lifetime of learning with a lifelong love of the natural world. Blending ecology, science, storytelling, and gentle graybeard humor, he reminds readers that environmental stewardship begins not with fear or guilt, but with gratitude, humility, and the simple habit of noticing.

When he isn't writing or speaking, he can often be found walking holding his wife's hand, photographing forgotten places, reading far too many books at the same time, watching birds that seem entirely unimpressed by human schedules, or standing in his garden wondering how the squirrels continue to outsmart him.

He remains convinced that curiosity improves with age, that every neighborhood contains a thousand unnoticed miracles, and that the world has never been ordinary.

It has simply been waiting for us to look a little closer.

www.ingramcontent.com/pod-product-compliance
Lightning Source LLC
Chambersburg PA
CBHW041259210926
51847CB00009B/73
9781956688856